To the second-year people

Keith County
JOURNAL

by John Janovy, Jr.

With Illustrations
by the Author

University of Nebraska Press
Lincoln and London

© 1978 by John Janovy Jr. Published by arrangement with
St. Martin's Press, Incorporated.
Preface © 1996 by the University of Nebraska Press
All rights reserved
Manufactured in the United States of America

♾ The paper in this book meets the minimum requirements of American
National Standard for Information Sciences—Permanence of Paper for Printed
Library Materials, ANSI Z39.48-1984.

First Bison Books printing: 1996

Library of Congress Cataloging-in-Publication Data
Janovy, John.
Keith County journal / by John Janovy, Jr.; with illustrations by the author.
p.   cm.
Originally published: New York: St. Martin's Press, c1978. With a new pref.
ISBN 0-8032-7588-9 (pbk.: alk. paper)
1. Natural history—Nebraska—Keith County.   I. Title.
QH105.N2J35   1996
508.782'89—dc20
95-39487   CIP

Lark Sparrow on page 5 is from the collection of Karen Janovy.

Practice Curlews on page 86 and frontispiece is from the collection of Jeanette
Dennis.

Great Blue Heron on page 108 is from the collection of Terry Stentz.

Western Grebe on page 174 is from the collection of Peggy Zalucha.

All other illustrations are from the collection of the author.

Preface

Keith County Journal was started in 1976. There are two versions of its origins. In the first, I take some bird paintings out to Jon Farrar at *NEBRASKAland Magazine* and ask whether he'd be interested in publishing them. He says yes, but asks for a paragraph about each one; the paragraphs turn into chapters which then turn into a book. In the second version, I'm lying in bed before dawn one morning wondering why anyone would put his whole sense of self-worth in the hands of a Nebraska state agency, decide no one should do that, then get up and write the first sentence: *Termite country, Keith County, is a windswept grassland.* Then I get up every day at 4:00 A.M. for the next year and finish the book. Both versions are essentially true, and only slightly embellished.

Rumor also has it that *Keith County Journal* was rejected twenty-two times before it was accepted for publication by St. Martin's Press. That rumor also is true. The rejection comments ranged from "beautifully written but not commercial" to "why do you waste your postage sending us something that doesn't turn us on?" I've put all those rejection letters to great use; budding writers of all ages love to hear them. In the year following the publication of *KCJ*, I received many "fan" letters about the book from people all over the country, phone calls from strangers, requests for speaking engagements, and solicitations from agents. Periodically, people even called me "Keith." Few of those things happened with any of my subsequent several books, all of which I personally consider equally important and mostly better written than *KCJ*. *Keith County Journal* has been a hard act to follow; there is something special about it, and I have no idea what that "something" is, unless it's the raw authenticity of the prose.

Basically, this book consists of the conversations I had over a period of several years with Vietnam War–era university students interested in biology. This is how we talked about nature, these are the views we expressed to one another, these are the acts that we did routinely in the course of also doing our real science, and these are the "second-year people" to whom the book is dedicated. The conversations, translated into formal chapters, give readers a peek into the minds of old-time parasitologists, at least, and perhaps into the mind of a type of scientific explorer that most of us in the profession believe is now endangered, if not extinct. When living, such beasts went to strange new places, even though the places may have been in the vacant lot next door; saw biological problems everywhere, even in the tiniest of organisms' lives; freely expressed the wonder in their work without every worrying whether they'd received a federal grant or graced their institution with a lucrative patent; and stepped into the classroom determined to make the world safe for the ideas of like-minded young people.

If academia is hiring, tenuring, and promoting such explorers and teachers today, then it's a well-kept secret. And the supply of "second-year people" also is dwindling rapidly. Twenty-five years ago they were everywhere; today you have to search diligently for a handful, and even when you find one, you end up wishing he or she was a little more disrespectful—in a productive way, of course—or a little bit more emotional, more given to creative acts done easily, and shamelessly, in public. I don't believe university students are any less creative than they were in the late 1960s and early 1970s, but I do believe they're less inclined to act creatively in public. Even at the time *KCJ* was being written, I'd come to view science as a relatively easy profession; papers in learned journals were not very difficult to produce then, and they're not particularly difficult to produce today. It's the "arts" part of "arts and sciences" that's so hard to find in a young scientist, or to develop in a student who's been told that he or she "doesn't need it for pre-med."

Yet over the past thirty-plus years of studying living organisms, I've come to view the artistic components of our thinking as essential to the scientific. Why? Because insight, elegant and testable hypotheses, clever experimental designs, patience with seemingly intractable materials (read "beetles are really stupid"), and inspired teaching are all products of our artistic sensibilities, our creative efforts. You can't teach a person how to ask insightful questions or how to be an inspiring teacher in the same way you can teach a person to do a chemical analysis. No, these traits are developed by the same methods one learns to play a musical instrument: through repetition, encouragement, help, patience, and a growing satisfaction with the increasingly rich quality of one's noise. The University of Nebraska's Cedar Point Biological Station—where *Keith County Journal* takes place—still is an excellent place to learn the art of doing biology. There are as many large problems lying about in the prairies now as there were in 1975, and these problems require just as much patience and insight as the ones we tried to solve two decades ago. And, I keep looking for more "second-year people" to tackle them.

The foreword to the original edition of *KCJ* talks about gifts, which were mainly opportunities, as well as about how to behave when handed something of value. For the past three decades, the University of Nebraska has been a truly wonderful place to pursue my career. I would like to thank the parents of this state for sending me a steady stream of truly bright, capable, and interesting young minds. There is no greater privilege than to spend your professional life surrounded by university students. Yes, several hundred of them went to Keith County, too, and yes, they got very dirty and tangled up in mud and vegetation and wild animals that might have been too small to see without a microscope. But everything washed off, at least off the outside.

John Janovy Jr.
September 1995

Foreword

THERE WAS A TIME shortly after Cindy was born, when a woman appeared at our door with a baby gift. The gift was, we felt, all out of proportion to the event, a gift well beyond that called for from a faculty wife fulfilling a social obligation to a departmental graduate student. Our reaction was consistent with our feelings, but our reaction was not passed off lightly, not at all. Our reaction elicited a chastisement, and Karen was chastised most severely for not accepting a gift gracefully, no matter what the gift. The woman's lecture centered around the lesson to be learned from gracefully accepting something of value, since failure to do so meant the giver's own feelings were not given fullest consideration in the response. That lesson has been well learned, for I and those close to me have been the recipients of untold gifts, whose value increases daily. The donors have very often been people who did not realize at the time what they were giving. The gifts have been those of time, talents, visions, performances; but most of all they have been the gifts of opportunity, and it is these last ones I need to talk about now.

The people in these stories are true and real, and there are people behind these stories who are also true and real but whose names are not mentioned. There is Dan Harlow, the friend who showed up at my house nightly with a bottle of Scotch to carry on a year's conversation about the tapeworms of sandpipers. His gift was that of the field of parasitology. There is John Teague

Self, the professor, whose major gift was that of strength and conviction in the face of adversity; guts and perspective. I have the feeling that much of what he gave fell from my shoulders onto the ground and was lost. He also finished off, in the way only he can, the process that Dan Harlow started. There are two students, Pierre Daggett and Joan Decker, who lived Agassiz's advice to study nature instead of books and reaped both the consequences and rewards, for in this day and age the study of books is a necessary part of the ritual of becoming a scientist. There is Leslie Stauber, now dead, who was able to say with the utmost sincerity that "somebody has to study that." There is Gary Hergenrader, whose quickness and insight gave me the opportunity to go to Ogallala in the beginning. And above all there are those people in Keith County, Mrs. Goodall, Dr. and Mrs. Gainsforth, whose insight and perspective gave me in turn the facilities and the physical means to look at Lewellen. There is Bill Current, a graduate student whose skill with the electron microscope revealed the dance of the cells within that one creature whose existence in turn justifies these chronicles as we live them in the South Platte. These people are the kind whose lives and styles have together provided opportunity; and in some cases this opportunity is obvious and tangible, but in other cases it is nothing more (I say nothing more!) than freedom from little concerns, freedom from stultifying lines of thought, freedom from the life elements that if taken too seriously, block greener paths of the mind.

There are good friends in these stories, and there are people who I consider exceptionally good friends but have never met. You will recognize yourselves, those who are good friends in the regular sense, and it is my sincere hope that those of you whom I have never met will also recognize yourselves as good friends, for in every case you have opened doors that I have stepped through, usually in muddy jeans and usually with a bunch of students close behind, though sometimes with a bunch of students up in front. If you see yourself in these pages but are unable to recognize me on the street, please rest assured that I am indeed, according to that lesson taught by that lady soon after Cindy was born, trying gracefully to accept your gifts of opportunity.

Perhaps I'm an opportunist, and there have been times when in Nebraska that is a compliment only if one is a running back or a member of the defensive secondary. I have been called overly confident, and again there have been times when in Nebraska that is a compliment only if one takes the snap from center. There have been times late at night when a wife has shaken her head, one more time for the last eighteen years, at the idea her husband would have the confidence to actually try to write a book about what he found by being an opportunist gracefully accepting the gifts of opportunity from friends. This is thus a time to thank those who have already read this statement of what there is to learn *from* nature rather than *about* nature and have seen something in it. This is a time to thank Pam Rhoten, who by keeping my desk clear and her typewriter busy, provided the time to wash that glassware, a love-joy task that generated many productive thoughts, as you will read later on. This is also a time to thank and mention the name of one of those, Dr. Richard Boohar, who acting out of total confidence in his own judgment, obligated his editorial expertise for the measly price of a couple of watercolor paintings of wrens!

J. JANOVY, JR., *March 1, 1978*

Contents

Keith County
JOURNAL

Termite Country

TERMITE COUNTRY, Keith County, is a windswept grassland. A life on the plains conditions a lad, produces an alertness for life in windswept grasslands, produces a caution in that plains lad looking for life, and this conditioning and caution in turn produce a feeling that the plains are not to be written off so readily as a mountain man is likely to do. A mountain man's appreciation for the Rockies and High Sierra is conveyed to the public, and the public is easily led to those national parks and places of great visual beauty, of breathtaking scenery, of sculptured sites, hard ridges and tall pines. The plains lad always comes off second best, it seems, in confrontations with mountain men, but with every second-best experience also comes a conviction that there is more to be seen on the plains than the mountain man is willing to admit, that maybe, just maybe, on those windswept grasslands there are lessons to be learned, sights to be seen, subtleties that sharpen the eyes and ears, and most of all relationships between living things that are worthy of the attention of a boy. A plains lad went to the mountains one time, and spent some other time in the pines, and both times he caught claustrophobia. His eyes looked for a flat horizon and saw none except in the back of his mind, and those horizons were beyond the mountains and pines at the point where the plains began again.

A history of society's life on the plains is a history of pioneers, of survival in what at first appeared a harsh and unyielding environment, of determination, of dependence upon one's neighbors and upon use of the soil and its products and inhabitants, of energy from fires built of buffalo chips, of soddies tunneled into the land itself, of families giving rise to families, of dealings with the natives, of searches for water. The history of society's life on the plains continues to be written, although often by acts performed on this land with no clear idea of their consequences, by diversions of prairie rivers, by irrigation canals, by the tapping of aquifers, by human plans and activities that only marginally take into account the extent to which others, others of less obvious lives, may in fact require conditions of the plains as they existed in the time of the mastodons.

The plains have been pierced in places, although those places are south of Keith County, and as a result the plains have spurted the gas and oil that propel a station wagon down Interstate 80 toward the Ogallala exit ramp. The ramp is the jumping-off place for termite country and the bluffs of Cedar Point. We climbed those bluffs the first year, for they invite one to climb; and the country looked snaky, so we took along our herp stick, made from an old hoe with the blade cut off. We were looking for life that first year, and had yet to demand of ourselves that we *do* something with the life we found, so that our daily activities were little more than a ramble through the grasslands, a ramble, however, increasingly punctuated with discoveries. Some of what we saw we had already known but had long since forgotten, and some of what we saw we had already known and easily remembered. But most of all, we saw things we didn't really understand, and we asked ourselves what there was that could be learned about them. The second year we also came to Keith County, once again climbed the bluffs and stood among the cow pies and hard grass, but by this time we had learned something about the nature of Keith County. We half-consciously did what any biologist with a herp stick up on the grassland of Keith County would do: we flipped over cow pies as we walked among the grasshoppers and lark sparrows, and it wasn't long before we found termites. The boy in the biologist remembers saying to himself at the time, The termite-under-the-cow-pie trick is one to be filed away for future reference!

Termite Country

There is very little wood in Keith County, although the junipers do drop occasional small limbs, and the ranchers drop occasional boards, often soaked with creosote years before and hardly fit food for termites. Cattle drop plenty into the grasslands and stones of Keith County, however, and beneath these drops lie the termite colonies. I mentioned this to the landowner. He had about twenty buildings, wooden, on the property, and he looked up startled, almost a double take.

Before it was cattle country, Keith County was bison country, and before that mastodon country, and it's hard to believe the mastodon and bison herds dropped less on the prairie than the cattle do today. It is easy to imagine the termites beneath piles of mastodon dung. It is easy to imagine the termites as oblivious of human progress; oblivious of alterations in land use, development of the region as a recreational and tourist area; oblivious of the land-acquisition schemes of local ranchers, the failures of some to manage, the bankruptcies, the purchases of mile after mile of rolling sandhills grassland, the purchases of whole sides of highways between towns, the irrigation canals, diversion dams, and large earthen dams that threaten miles of river communities when the wind blows from the west for several days, as it often does.

Down in the valleys between the bluffs there may be some termites, in the leaf litter beneath the junipers where it is dark, where the late spring rains soak instantly into the sand and any self-respecting termite might very well tunnel clear down into the aquifer if necessary to maintain the 100 percent humidity characteristic of termite nests. In the valleys the common birds nest, and when but a few hours out of the egg get their first cases of malaria. Among the rocks of the bluff face lie the bull snakes, in equilibrium with the white-footed mice that form their food. Magpies careen up and down the canyons. The North Platte River roars in the night. In the cattails, marsh wrens buzz away the days inches above the trillions of conical snails. This is the Keith County prairie. There is reality down in the valleys.

Reality stops at the rim of the canyon, just as reality must have stopped at about Saint Joseph, Missouri, in years past. The human

[3]

in the canyon climbs the face of the bluff and turns over a cow pie. There are the termites. The human makes his way across the prairie and turns over still more. Under almost every one of these are the termites. The human rubs his forehead—termites are science fiction anyway, their normal lives strain credibility. To live on top of the bluffs is to live in the teeth of the elements. Here the termite with its total dependence on moisture and darkness lives in that part of Keith County where darkness and moisture only marginally exist themselves. There must be some mediator, and of course there is, in the form of herbivore dung. I use this term *herbivore dung* very much on purpose, rather than *cow pie*, for the great herds that stomped down this hard grass before man arrived must also have supported a termite population in what we now call Keith County, and this knowledge tends to put our own so-recent pioneers' dependence upon tunnels and herbivore dung into some kind of perspective. The human sitting in a soddie cooking over a fire of buffalo chips was really doing nothing more than these termites had done for millions of years, moving into a seemingly harsh plains environment, digging into the ground itself, extracting the energy from herbivore dung. It makes one think long and hard about how different we might actually be from these other inhabitants of Keith County.

Termites are colonial insects, however, and as such are supposed to possess a rigid caste system, a division of labor, a non-human predestination for roles that an individual has little chance of changing. A biologist tells us that in Keith County *Reticulitermes tibialis* is the animal but the colony is the evolving unit, and in that statement is implied total loss of the individual's freedom, potential, and developmental possibilities. There is no such thing in nature as a single termite, just as there is no such thing as a single honeybee or a single cliff swallow. To say the individual is subordinate to the group is to understate the case by an order of magnitude. The group *is* the organism, the individual is one of the organism's cells. The species *R. tibialis* occurs on the face of the earth as a population of evolving units, not as a population of individual animals any more than the population of cattle on the face of the earth occurs as individual cow cells.

Lark Sparrow

". . . we flipped over cow pies as we walked among the grasshoppers and lark sparrows, and it wasn't long before we found termites."

[5]

The name *R. tibialis* requires the coordination of the colony just as the word *cow* requires the coordination of the cow's cells.

The evolution of colonies, of this life-style we have described, is a matter of millions of years. It is not an event that occurs over the Labor Day weekend. It is an event that requires a colony reproduce in a manner analogous to that of a lower animal. It is also an event that assumes—and I probably need to say that word again out in Keith County—that assumes the continued existence of an energy supply in the form of herbivore dung.

The termite colony reproduces asexually as well as sexually: asexually it buds. The subterranean passageways and tunnels extend through the soft soil until finally the lines of communication with the parent colony are stretched to the point of breaking, and a new colony is formed at the end of some passageway. At the end of the tunnel the pioneers and explorers undergo a biochemical transformation, and on their own in the wilderness assume roles they would never have assumed in the bustling and structured metropolis. Workers' hormone production becomes unsuppressed for a select few (perhaps preconditioned from birth and predestined to be inducible and seducible?), and these few become supplementary reproductives, or supplementaries. The supplementaries will reproduce sexually, and they may simply be the first to become supplementarient. With their newfound ability to supplement, they may suppress the development of further supplementaries. Supplementariness is a virtue only in the colonies. Somerset Maugham is full of supplementary virtue in the colonies. It is extremely doubtful that a supplementary would return to the parent colony; the tunnel has fallen into disrepair. Before many years the link between the parent colony and its now supplementary-equipped daughter will be lost. New workers, the products of supplementaries, have no biochemical ties to the parent colony, only to the supplementaries. And the colony has reproduced. Life thus goes on in Keith County, and this is a long-term activity, a million-year activity, that assumes the continuous energy supply of cow pies. The termite popuation of Keith County is evolving no contingency plan in the event that one day there will be no more herbivore dung.

Termite Country

The life of a Keith County herbivore is nothing like that of the Keith County termite, for this is cattle country. To say the cattleman "loves his cattle" is to cause the initiated to blink. *Love* is the wrong word, so inadequate, so correct in intent but so far off base in context as to make one blink. To the cattleman there are cattle just as there are sunrise, sunset, hard grass and pickup trucks with rifles cradled in gunracks across the back window. One senses a vast and virtually overwhelming empathy for cattle; the cattleman knows and knows well these animals give him his pickup trucks and gunracks, and he knows and knows well that they will be systematically killed, one by one, as efficiently as possible by a bunch of guys wielding electric guns and long sharpening-worn knives.

One wonders if the cattleman makes the visit inside the packing plant, where an animal is reduced to meat in a matter of seconds. There is some writhing and kicking at death, but within seconds the animal is strung up by its hind legs and the skull removed. The sight of a cow or steer body, sometimes with its patterns of brown, black and white, hanging by the hind legs with a great flap of hide where the skull used to be, is one of the single most uncattle sights within the walls of that packing plant. A steer in this condition has in one act been totally de-steered in every sense of the word. What happens from then on is trivial: the animal is cut up. The removal of the skull is an event in the life of every beef animal, and it is the event that in a few brief moments transforms the creature into a spectre, after which it is all right for it to become steak and roast.

In the field, however, cattle appear dumb to the outsider. They empty their bowels seemingly continuously. They stand at the fence and look, or else skitter up the ravines. Someone has tagged every one with a red ear-tag. They wheel on their hind legs and stomp off through sunflowers higher than your head. There is an occasional great bull, and there are many calves. They chew; they seem to chew more than they actually eat. One never sees them on top of the bluffs, however, and one then wonders how all that manure got up there. We have observed cattle for two years now, at least in the summers, and have yet to see one on

top of the bluffs. Nor have we seen one in any kind of relationship with a termite, no consultations, no actions on the part of
cattle that would lead us to believe a steer is responding to the
insect; yet there is a world of steer manure up on the windswept
grasslands. No one carts cow pies up to the bluffs, so at some time
in an annual cycle there are cattle up there, benevolently concen-
trating the cellulose grass-stems for the termites.

To the termite above the rim the emptying of cattle bowels is
a matter of life and death. Here is the cellulose to feed the intestinal protozoan symbionts that feed and live with the termite.
Reticulitermes tibialis has an intestine full of other animals, and
unless a person has seen those animals inside a termite intestine,
there is no way to accurately describe them. They are exceedingly
complicated creatures to be living inside a termite, and one must
ask why they possess such ornaments and finery for a life im-
mersed in the digestive soup of a termite intestine. They are
twisted and hairy, they move in great ripples, they form a com-
pact blob when the termite intestine is ripped open, they come
in many sizes and species, and they are quickly killed by the very
oxygen that we breathe so readily. They are also full of cow
manure. In the microscope manure looks like splinters and crys-
tals inside a one-celled animal inside a termite. The termite has
reduced the dung to splinters small enough to be taken into the
symbionts, and these one-celled animals have scarfed up the
splinters in their anaerobic world. The splinters are cellulose.
Cellulose is made of the same stuff that starch is, a simple sugar,
but there is a world of difference between simple sugars arranged
as starch and simple sugars arranged as cellulose.

The human sits on a chair and eats a french fry, and that is the
difference between starch and cellulose. Both are synthesized by
plants, both are made of the simple sugar glucose, both are made
of long chains of glucose molecules linked together, sometimes
branching, both can be burned in a fire for warmth; but only
starch can be digested by the human. The human builds furniture
and shelving out of cellulose, studs in a house under construc-
tion, boats, sculpture, laminated bowls and trays, ceremonial
masks, fence posts, and railroad ties. The vast sociological differ-

ence between starch and cellulose is centered in the links of the chains of glucose molecules. Link the molecules one way, and the chain is the major source of energy for most animal life on the face of the earth and the chain is what the plant kingdom builds to fuel the animal kingdom. Link the molecules the other way and the chain is indigestible, the chain is hard, bundles of chain become rigid, can be sawed, planed, drilled, caulked. Link the molecules this other way and the chain is wood. The cellulose part of the steer's diet passes through the intestine in large almost unchanged quantities. Only the termite eats cellulose, and in fact the termite does not really eat cellulose, it simply prepares it for its intestinal symbionts, who actually do the digesting.

One would almost think that this ability to break the cellulose link should give these protozoa the right to evolve their finery. Consider what would befall the man who could break this cellulose link, easily and cheaply: he would have a patent on the mechanism for easily and cheaply converting wood into food. In today's market, he might as well have a patent on a mechanism for spinning straw into gold. Such a man would be covered with finery compared to his relatives. So it is with the one-celled animals, the protozoa, of the termite intestine. They are large, often corpulent, with stomach-equivalents crammed with cellulose splinters; flagella fall from their shoulders as ribbons and sashes and waist-length hair; they glide with slow dignity, undulate, like important individuals equipped with unique powers, which they are.

Inside the termite there is none of the helter-skelter and frenetic searching of lower protozoa, the batting off the walls, the heart attack pace of test tube life. Inside the termite there is slow dignity without oxygen. To the protozoan, the dignity is inherited. The ability to convert wood into food is inherited and is based on the ability to synthesize a protein molecule, an enzyme, which in turn acts to break apart the links of the glucose chain. The inherited ability is provided by a gene. In the gene is a set of instructions that the one-celled animal uses to construct the protein that breaks apart the chain of glucose molecules. The instructions are information and the information is unique. The one-celled animal has information and instructions that others do

[9]

not, and it uses those instructions and that information to perform an act that not even the most technical society, a society that can send men to the moon, can perform. There is no clearer demonstration that information is power. This biological talent also supports the pest extermination business in the United States of America. The employees of Orkin support their families on the information that is contained in the genes inside one-celled animals that live inside termites, and that give the one-celled animals unique transformational powers.

The unique transformational powers are those that are required for the extraction of energy from cow pie cellulose, and as it turns out not only the protozoan symbionts but also the termites are users of this energy. It takes energy to carry out tasks and to an animal energy is currency. One must spend energy in order to get energy, and in living systems it is hoped that the energy received will be more than the energy spent. It better be. The energy a termite intestinal protozoan spends to make its enzyme is the energy that a one-celled animal spends in order to get energy. Since the termite is the fortunate recipient of some of the cellulose energy released by these protozoa, the energy the termite spends to maintain a life and home for the protozoan is the energy the *termite* spends to get energy. There is a satisfaction, a sense of wonder and relief that things are working properly, that comes from just thinking about the mutualistic relationship between termite and protozoa. It is only when the steer is considered that there is produced some intellectual discomfort, some sense, feel, of a potential impending breakdown in the web of interrelationships that permits a life on the windswept grasslands of Keith County.

This vague worry, this sense of possible catastrophe, is derived from our knowledge of not only the steer's role in the interrelationship web, but also, and probably most strongly, from our sense of the steer's total lack of awareness of its own role. While the termite-symbiont and colony units have committed their lives to an activity conducted by herbivores, the specific activity of eliminating wastes, the herbivore is in fact an unwitting and fortuitous participant in the lives of the insects and protozoa. The cow does not know the relationship that exists beneath its feet, and the biologist in the boy immediately begins to wonder what kind of elements might go into a cattle decision to defecate somewhere else. Such thoughts drive one quickly to action, for one also wonders how many of these kinds of relationships one participates in oneself, and so a person stomps into the fields of Keith County to learn what else is there, what other lesson is there to be ranked upward and included in human affairs at decision times.

The termites started this all, they did, with their dependence upon the herbivore, their heritage of dependence upon prehistoric herbivores with different roles than beef cattle, upon prehistoric herbivores that answered to a different call than do beef cattle. But then in the second year it became obvious that we had only just started when we turned over cow pies and found termites, and that although the termites had many things to say to us, there were others, such as snails and swallows, that had *their* things to say, and those things were also things to be filed away by a plains lad for future reference.

2

The Snail King

STAGNICOLA ELODES is the snail king of Keystone Lake. The animal is slightly less robust than the terminal joint of an average human little finger, and it thrives in astronomical numbers just below the surface of Keystone Lake cattail marsh. It's a large conical snail for the area, and fit food for any number of higher as well as lower species. The words *Stagnicola elodes* translate from an ancient language into modern English as "one who dwells in the stagnant marsh," and few animals have ever been named so truly. Snails move with dignity and detachment along the submerged and rotted stems of last year's cattails, in the center of the marsh, where it's warm and still.

Keystone Lake lies just downstream from Lake Ogallala, and is fed from Lake Ogallala by channels through the marsh. Lake Ogallala is fed from Lake McConaughy, or "Big Mac" as the natives call it, which is twenty-two miles long and four or five miles wide. Big Mac is a hundred feet deep and bitter cold except in the shallows of Arthur Bay or at the crystal swimming beaches along the dam. The spillway draws water from the bottom of Big Mac and dumps it into Lake Ogallala. The water is bitter cold. Lake Ogallala is bitter cold. Keystone Lake is bitter cold. The North Platte River that drains Keystone Lake is bitter cold. Except in the cattail marsh; there it's warm. *Stagnicola elodes* lives in warm, surrounded by bitter cold.

The eastern end of Keystone Lake is a diversion dam, with a

set of gates on the south. Keystone water can be allowed to flow down the North Platte River or it can be diverted into the Sutherland Irrigation Canal. James Michener wrote about this land and about structures like the Sutherland Canal. The book is entitled *Centennial*. The canal is named after the town of Sutherland, which is the closest town to the former home of Charles Simants, a man convicted of slaying an entire family of six neighbors. The Simants case went to the United States Supreme Court, not because of Simants, but because of the judge who banned press coverage of the preliminary hearings following Simants's arrest. The press has over the years dutifully reported the arrests and convictions of accused and convicted mass murderers. Mass murder convictions are routinely converted into books, including such masterful books as Truman Capote's *In Cold Blood*. Not a few multiple murders occur on the plains. All are covered by the press to the best of its ability; and when the ability is curtailed, then the press covers the curtailment of the press's ability to cover multiple murder. It is not yet clear what humanity has learned from press coverage of multiple murder. The press coverage of curtailment of press coverage of multiple murders may in fact yield more useful information for humanity than study of the murders themselves. To date there has been no curtailment of press coverage of the death of *Stagnicola elodes*, and no Supreme Court case to test the validity or constitutionality of such coverage has ever been heard. There are those—a few—who feel that press coverage of the death of *S. elodes* might ultimately be of more benefit to the world's human population than would press coverage of Charles Simants's preliminary hearings.

In June the land gets thirsty and the engineers give it a drink from Big Mac. The spillway is opened and tons of ice-cold water roar into Lake Ogallala, sweep into Keystone Lake, through the cattail marsh, and over the diversion dam and into both the North Platte River and the Sutherland Canal. The accumulated flotsam and debris of a Sandhills winter is ripped loose from the marsh and sent down the river and canal, and the level of the lakes, river, and canal is elevated a foot or two. With the debris goes *Stagnicola elodes*. The catharsis lasts for weeks, and for weeks

Stagnicola elodes is jerked from its warm home in the cattails and sent careening down the bitter-cold river.

Awaking early, the human wanders down to Keystone, across the wooden bridge and onto the large sand bar with its boat dock. The water is up; during the night the engineers have opened the gates. The human frowns and looks up and down the shore; plants that were five feet from the water are now submerged, there is a current through the lake, and pads of algae and bits of cattail stem are moving past the dock. Scratching his arm and yawning, the human turns back to breakfast but stops. Something has registered on his subconscious. He looks again at the lake, closely this time. On every pad of algae, on every bit of cattail stem, there is a snail. Sometimes there are several snails. The human stands on the dock and counts snails, *Stagnicola elodes*. In a minute fourteen *S. elodes* float by. In another minute sixteen *S. elodes* float by. This has been going on all night, since the gates were opened. The lake at this point is about a half-mile wide, but the man has watched a strip of only about six feet. There is debris scattered across the lake, all of it moving downstream at fifteen snails per minute. The gates are left open all through June and July and August. On an average day over nine million snails are sent downstream to Sutherland and North Platte, in an average summer, nearly a billion. They don't come from the bottom of Big Mac; they come from the cattail marshes of Lake Ogallala and Keystone Lake.

The man skips breakfast and moves purposefully into the cattails. The water is bitter cold in the channel itself, but warm and shallow, soaking through the jeans, in the cattails. *Stagnicola elodes* is still there. Each place the cattails are parted and the fetid water examined, there are the snails. They have not gone. But the pads of algae are still carrying snails down to North Platte, fifteen per minute. The human leaves the cattails just as *S. elodes* leaves the cattails: through the channel and open water. The human winces as he steps into the channel. One foot is still in the cattails, and it is warm and the mud is above his ankles. The other foot is in the channel, and the bitter cold water is up to the knee. It is early and cool, and the water gives the human a sense of headache. Still, the human is warm-blooded, possessed with biochemical mechanisms and clothing to maintain the body at thirty-seven degrees centigrade regardless of the immediate environment. Soon there will be coffee, maybe a jacket. There is no coffee for *S. elodes*, but nine million a day do the same thing the human is doing, stepping out of the marsh, leaving their warm home into what must be for them the most bitter of wilderness, on their way to North Platte. It is difficult to imagine the shock and stress of *S. elodes* out in the open water of Keystone. It is difficult to imagine a less hospitable environment than the one that exists a foot away from the cattails, the one into which *S. elodes* is thrust at the rate of nine million a day for three months of the year.

At the east end of Keystone Lake is a concrete diversion dam and gates to the Sutherland Canal. East of the dam, installed into the North Platte River, is a series of rock dams, each doing nothing more than breaking up the river, creating a series of rapids. The state record rainbow trout is said to have come from one of these pools below the diversion dam. Once a loon rode the current through one of these pools for the better part of a day, teasing the photographers. The rocks of these rapids are covered with blackfly larvae and pupae. They are also covered with *Stagnicola elodes* eggs, washed down from the marshes of Keystone and Ogallala.

Stagnicola elodes lays a string of eggs in a gelatinous case. Under the microscope the string is beautifully coiled within the mucous, the tiny brown eggs forming a spiral. When the rocks are lifted, out of the rock dams and rapids, they are seen to be covered with these gelatinous egg masses. More rocks are lifted out of the rapids, and all are covered with egg masses. The available rock surface of the entire dam is covered with *Stagnicola* eggs. There are no adults in the rock dam, no cattails, and the eggs can only have come from marshes nearly a mile away. The spillway does not discriminate against the young; it is simply that the young, while still eggs, have no ability to detach themselves from the rocks and raft down the river to perhaps a more hospitable clime. They are caught in the North Platte, stuck to the rocks. There is no way to count the snail eggs in these rock dams. They must outnumber the floating adults a hundred to one. It was intended that these eggs incubate in the cattails, in the warm and stagnant water below the wind. Their metabolic machinery, like that of the adults, must be adjusted by the experience of thousands of years of *Stagnicola elodes* generations to those conditions of oxygen, temperature, dissolved and rotting organic material, methane, one finds in the marsh. Their ability to develop in the rock-dam white water has not been determined, but no sensible biologist, attempting to raise *Stagnicola elodes* from eggs, would ever incubate those eggs under rock-dam conditions.

Animals are generally assumed to invade watersheds from *downstream*. There is no date recorded for the first occurrence of *S. elodes* in the Missouri River, or the Platte River, or the North Platte River. There is no record of where it came from. There are places along the Platte where cattails exist, although few places the equivalent of Keystone, and one must suspect there are *S. elodes* in these other places. Indeed, just below the rock dams are a few cattails along the side of the North Platte, and in these cattails, only a few square yards of marsh, are *Stagnicola*. It is not clear whether this population came from above the diversion dam, from Keystone, or from below, up the river. It is clear, however, that millions of snails and snail eggs go *down* the river from the Keystone marsh.

The initial trip up river for the species must have been difficult. History and fiction both tell us of the human difficulties of traveling upriver through the Missouri drainage, yet the human is marvelously equipped for exploration compared to an adult *S. elodes*. In history the human comes equipped with flatboat, Indian guides, traps, rifles, companions, determination, canoes, leather outer garments, and beads for trading. Out of that list, *Stagnicola elodes* can at the most have only determination, and that is only the determination that is inherent in every animal species. There is a shell, of course, and a tough muscular foot to carry it, but shells and tough feet are relative things. A heavy protective shell and a muscular foot to *Stagnicola* is little more than a tasty tidbit to a coot. It is difficult to imagine that at any time in geological history *Stagnicola elodes* moved upstream at the rate of nine million a day plus a hundred times that many eggs.

History is also full of human colonies established upriver, no sooner were they established than the colonies and outposts began communication downriver, usually at a considerably higher rate than the initial colonization effort. So it is with *Stagnicola elodes*: the major population established in Keystone, the road upstream blocked temporarily by hundred-foot-high Kingsley Dam, communication now begins downstream at the rate of nine million adults plus an undetermined number of eggs every day.

Communication to a human involves the transfer of information. Communication in a commercial sense, by an upstream outpost, means transport of raw materials. Raw materials, furs, minerals, precious stones and metals are extracted from the upstream site and sent pell-mell downstream to be converted into new things. Information and raw materals. Information as raw material? Perhaps in the case of human outposts, information can serve as raw material. Certainly in the case of the snail king information is the raw material, and it is sent downstream from the Keystone outpost, established in the recent geological past, at the rate of millions of bits a day for days on end.

Charles Darwin in the middle 1800s, and later Theodosious Dhobzhansky and Ernst Mayr, have told us what nature does with raw materials and information sent from an upstream outpost through an inhospitable environment. Nature makes new species of snail kings. The philosophical question is whether *Stagnicola elodes*, or for that matter any other animal, has a built-in proclivity for exposure to inhospitable conditions, knowing, in the genetic sense, that such exposure guarantees the change required for survival in a changing environment. If so, then nine million snails a day are doing the very thing that will, in the long run, convert the *Stagnicola elodes* genetic stock into other species, related species, perhaps able to survive in the rushing Platte, perhaps requiring the cold, the high oxygen, the moving water. Within each cell of a snail, within each egg, there is a set of information as there is a set of information in each of us. The information is in the form of a DNA molecule, and

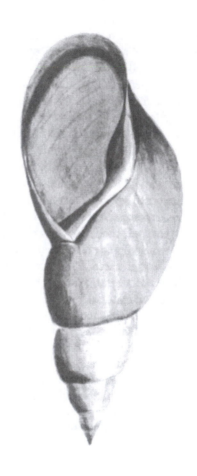

Stagnicola elodes

". . . nine million a day do the same thing the human is doing, stepping out of the marsh into the bitter cold, stepping out of their warm house and into what must be for them the most bitter of wilderness, on their way to North Platte."

for the average citizen DNA can be found in *Time, Newsweek,* and even the front page of the *Lincoln* Journal, where was dutifully reported the University of Michigan Board of Regents' six-to-one decision to allow certain kinds of genetic research to be conducted in the school's laboratories.

There is also an experiment of sorts, involving DNA, going on in Keystone Lake and the North Platte River. The Central Nebraska Public Power and Irrigation District, giving the land a drink from Big Mac, is in fact experimenting with *Stagnicola elodes* and in the long run will reveal whether *S. elodes* can survive by adapting to new and challenging situations. The stresses of the cold North Platte and the gushing rock-dam water will, over the course of time, demand and get some adjustment in the set of genetic information, the DNA, that now spells *Stagnicola elodes.* The snail king lives in unbelievable numbers in a warm marsh. The snail king and his eggs are lifted, in numbers just as unbelievable, out of the warm marsh into a cold and hostile environment. The forces of evolution will effect an adaptation at some time in the future. *Stagnicola elodes* will have coped.

For most animals on the face of the earth, death and stress are the normal circumstance. Many are born, whether they be snails or flies or badgers, but the majority die before they reproduce. This is the normal circumstance for the vast majority of animals, including *Stagnicola elodes.* They die primarily as youth, mainly through exposure to inhospitable conditions. On the other hand, the species, or more often a slightly modied species, survives.

The earth turns, and so day and night result; the earth orbits the sun, and so a year passes; the solar system moves with the arm of the galaxy; the galaxy is moving at almost the speed of light away from others; and the sun at the heart of our solar system is aging, one day to die itself. The forces of conservatism do not operate much in nature. The forces of progressivism, of change, of constant change as the only constant, operate continuously if slowly. *Stagnicola elodes,* through, in part, the courtesy of engineers, is living consistently with the forces of progress by dying in wholesale numbers in the wilderness of the open water, a foot away from home, in Keystone Lake.

It is possible, that if an election were held, a campaign conducted, opinions sought and expressed, aides solicited, organizers,
and supporters garnered, and all the activities organized as evidence
of some candidate's ability to organize and garner—that the voting
population of *Stagnicola elodes* in the Keystone Marsh would
decide as a body not to open the spillway this year. They might
feel that maintenance of the marsh was first priority, and that
in fact it was foolish to engage in activities, such as the opening
of the spillway, that not only were very expensive, in terms of
the welfare of the snails, but that exposed great numbers of the
voting population and their progeny to stressful conditions, and
from which no immediate economic benefit could be seen to arise.
One knows little of the inner workings of a snail brain, but history tells us that the voting results would be no different if the
stresses were intellectual rather than physical.

3

Pioneers

A Place in the Sun (American movie title).
". . . my island in the sun . . ." (Harry Belafonte).
Islands in the Stream (Hemingway book).
How Does One Get from Here to There? (children's book).
". . . a time to walk, a time to run . . . a time to live, a time to die. . . . take it easy, baby . . ." (The Doors).
". . . say to the wind as it takes you away, that's where I wanted to go today" (Jefferson Airplane).
"To every thing there is a season . . . a time to get . . . a time to keep" (The Bible, Ecclesiastes).
A time and a place for everything (trite saying).
A place for everything and everything in its place (familiar quotation).
"The bolt hole in the rocker arm shaft must be on the same side as the adjusting screw in the rocker arm" (*Motor* Auto Repair Manual, 1976, p. 1-539).
"Over the river and through the woods to . . . we go" (traditional song).
"We tried it and it didn't work" (Royce Ballinger describing the Texas-Ohio lizard transfer experiment).
". . . it's my life, ain't nobody gonna tell me how to live it" (Charlie Daniels).
". . . it was a very good year. . . . I did it my way . . ." (Frank Sinatra).

"Please reread your role statement" (young committee woman).
". . . love it or leave it" (bumper sticker).
"This is my time and this is my lab" (J. Janovy, Jr., Keith County).
"The water tanks around those wells are full of snails" (Mr. Chandler, north of Paxton).
"No, I didn't put them there" (Glen Drohman, greenhouse manager).

L IKE HELL you didn't put them there, I would have said a year earlier. I believed him that day, however, for I had seen the snails in those tanks on the Chandler property, many miles from the nearest *Stagnicola elodes* population; I had been to the rock dams downstream from Keystone Lake and panned up the eggs; I had waded ashore, MacArthur-style, near some beach house on the far south shore of Lake McConaughy and found a single strand of eggs, *Stagnicola* eggs; and I had stomped across a very dry Sandhills prairie with a world authority on such things, and he had told me tales about *Stagnicola elodes* that were outside my wildest speculations. The student with us had shaken his head and smiled; he had a way of doing that when, at times like this, my wildest speculations turned out not only to be true but in fact quite short of the truth.

The water is in a very small concrete pond, and the pond is drained every year because in this country water freezes to the bottom and cracks concrete ponds. The pond is in a small native plant garden between two buildings downtown in the capital city. The garden was built to honor the memory of an original botanist, and it is a beautiful garden. No one, and I say this with total conviction, no one, makes a garden like Glen Drohman. The garden is the size of your living room, but in that garden is the whole state, and the whole state undergoes the wondrous seasonal changes of the plains. The Sandhills are there, in part, a small part of Keith County nestled between two downtown buildings in the capital city. The pond is full of *Stagnicola elodes*.

"No, I didn't put them there," said Glen in response to my question. "Probably birds, don't you think?"

No, probably not birds, I thought, the kind of bird that transports *Stagnicola* is a coot, a duck, a great blue heron, maybe even the yellowlegs. The bird that frequents Glen Drohman's pond is a starling or an English sparrow. My conclusions about the uses of Glen Drohman's pond do not make the snails any less real; they are still there the next day and the next as I pass between these city buildings. They blew in from somewhere, but the extension of that thought leads quickly to the absurd, a vision of snail eggs all mixed up in the dust of some local small whirlwind, a vision of snails just stopping to rest on their way to other cattail marshes, a vision of mass migrations of snails of which humanity is totally unaware. There is no person, at least none that I know of, in this city, who goes around putting *Stagnicola* in outdoor ponds on purpose. My question to Glen was simply to clear the air for further consideration of the snail. Just wanted to make sure, just make sure, that he had not decided the animals were such a normal part of the Sandhills that he had put them there on purpose. Noticed accidentally, of course, and found in Glen's pond they were; but then after some time spent in the marsh one begins to subconsciously peruse every body of water for snails, even a glass of water served in a restaurant. It could well be that a snail find is nothing more than an expected result of a snail hunt, but then a snail hunt is really nothing more than an expected result of the first snail find. We will notice next year whether Glen's pond still rates *Stagnicola elodes* from some unknown source.

We are hunting snails now, simply hunting snails out there in Keith County, out where the water is, for our snails need some water, and we are simply going to write down what we find and where we find it. We might take some pictures to remind us months later of where we went. Hunt snails and write down where you find them.

One has to hunt snails for a year or two to get the feel of such an activity—the feel, the options, the approaches. They range from wading in the mud to wading in the mud. There are things that happen between muddy spots, however, such as the *Boston Whaler*. The pictures taken from the *Boston Whaler* that day

[23]

comprise one of the strangest sets of pictures I have ever seen, and a good many of them are my own. Generally they are pictures of where *Stagnicola elodes* is not, and they were taken that cold day out on Big Mac. *Stagnicola elodes* is where a worm lives, so that a worm has two places to live: *S. elodes* and the place that *S. elodes* lives. If you subtract from all places the place that *Stagnicola elodes* is not, and if you subtract from all times the times that *S. elodes* is not, then what remains is the time and place of *Stagnicola elodes*.

The snail king is in Keystone Lake, no doubt about that. But one walks easily upstream from Keystone and within a half mile runs into the dry sand mountain of Kingsley Dam. Struggling to the top of Kingsley Dam one looks out toward Lewellen over twenty-two miles of crystal water with not a cattail in sight. There are times late in the afternoon when you drive to near the end of Kingsley Dam, sit on a log and drink a beer and think about the day, looking toward Lewellen. You wonder about the snail king, you wonder if there is a time and a place for him in Lewellen, what he thinks and does about Kingsley Dam; you wonder if he is out there in Big Mac, anywhere; you wonder if he's pioneered beyond the earthen mountain to find his time and place in some cattail marsh upstream; you wonder if he's working his way around the barrier, forcing the issue further and further upstream, against the current, seeking places. Not many pioneers you read about ever worked their way downstream. They worked their way upstream until they came to the mountains. *Stagnicola elodes*, the "one who dwells in the stagnant marsh," is upstream to Keystone Lake, looking at the mountains of Kingsley Dam. We will see if he's made it across. It's also a great excuse to get out the *Boston Whaler* and spend the day on McConaughy. It's cold as hell, busting across the waves, and we are huddled in the *Whaler* searching the shore. Up a cove, there are some trees; looks like weeds in the water a quarter of a mile up the flooded canyon. We work the boat into the tributary, up into the shallows, and I wade ashore to find what I know is there: no snails. This is not the time and place of *Stagnicola*. I photograph the spot.

The south shore of McConaughy is what remains of the Brule

outcrop canyons, flooded by the waters back in the 1940s. The rocky tops of those bluffs form a craggy beach head, but there are pioneer cottonwoods and willows now up in those canyons rather than old-timer hackberry trees. The sand has washed out of somewhere and made small beaches below the rocky tops and sheer cliffs of the south shore. The seclusion of these beaches cannot be told. One can sail out of Arthur Bay across the lake to one of these beaches and go swimming, very alone. The south shore stretches west for twenty-two miles, and toward the west there are more trees, mud flats, brushy canyons, houses on the walls of those canyons, and places where a snail might be.

You have to look now, busting along through the walleye fishermen, for snail places and snail times, and you pull into cove after cove, every likely spot for fifteen miles of south shore, and you find no snails. You photograph every no-snail spot with care, for no-snail spots look just about like snail spots from a distance, from the middle of McConaughy. There are some white pelicans at one spot—one especially interesting and likely spot—and you photograph the pelicans. But there are no snails. The ritual is the same; you scream over the roar of the Evinrude, smiling, pointing, crashing through spray and waves of midlake, and your pilot wheels the *Whaler* inshore, cutting the engine as the backwash lifts the stern, the engine gurgling now and steadying into a trolling pace as you work up into the no-snail spot. That's all, he says, too shallow; and over the side you go with camera, white pan, jars, field bag, over the side into thigh-deep water, bitter cold, over the side wading after them babies! You pan. You look. You stomp through the water of some cove or upper end of some flooded canyon, and you find nothing. You photograph.

You will wonder later, sitting on that log in the blistering late-afternoon Keith County summer, why you did all that. You will wonder, sitting on that log with a late afternoon beer, looking toward Lewellen, why you spent that day wading after the places where *Stagnicola* was not. Late that summer day sitting on the log you will have your answer, for in one of those coves, the one with the beach house you can still see so vividly, perched on the side of that bluff, you found some eggs. They were in the pan. You had waded ashore, stopped at the weeds ankle-deep in the main lake; waves from the north wind were crashing into your legs, and you reached down, pulled a weed, and washed it in your pan. There was a single strand of eggs. Stuck to the weed they'd been. You washed every weed in the cove and found no more. A single strand of *Stagnicola elodes* eggs on the whole south shore of a twenty-two mile-long lake. Pioneers. The species is looking for its place and time. You wash the eggs back into the lake where you found them. A year from now there may be snails up in that cove, in some temporary pond separated from the main lake in that brush. On the other hand, that single strand of eggs may perish. The snail king tried it and it didn't work, he might say. If it didn't work, it's not going to be because of you. You photograph the place where the eggs were found. Later you're sitting on the log looking toward Lewellen and you know that fifteen miles down the south shore was a single strand of *Stagnicola elodes* eggs. Somebody made it across the mountains. You will look later in the Lewellen marshes, but you don't need to. You know the snail king is there. If a snail egg can make it fifteen miles along the south shore, then the species can make it to Lewellen.

The waves on a stock tank are not large, not nearly so large as those that crash against Kingsley Dam and bring sightseers out from town. The stock tank is metal, circular, almost waist-high and filled with water from the Chandler wells. We were thinking of buying some Chandler property, looking over the ground, assessing its value as a place to study biology. Simply study biology. Mr. Chandler had driven us the fifteen or twenty miles from his house to the property, driven us in his pickup, driven us very fast along slow roads. Later the director would do the same thing in his 1967 Chevy and come within an inch of killing us, missing a turn. Mr. Chandler missed no turns. He missed no turns out in the roadless sandhills of his property either, never slowing, finding those places on the two sections that he wanted to show us. The biology teacher made him stop a time or two, especially after he said that bit about the snails in the tanks.

Chandler stopped at his favorite spots, Chandler places on

Chandler property; this was Chandler time, and he really did not know then that we would be unable to meet his price. For all he knew, out there on Chandler property he was riding with the guy who would control what was now his ranch; for all he knew that day, he was looking at his places with the knowledge that his opportunities for doing that would be limited. There must be some set of clues, though I have no idea what they are, that tell a place-and-time person that another is also a place-and-time person. Mr. Chandler had his places on his property. They may not have been the places that would have been mine, had the property been mine, and indeed I saw many places that afternoon that had I owned the property would have been special. Chandler's places were special, were so obviously *his* places. He showed them to us with pride. No one knows what kind of thoughts were in his head those times he must have been on a particular hillside, twenty miles from his house, maybe with two feet of snow on the ground, maybe 108° and no rain for months.

There *were* some thoughts, however, some Chandler thoughts. I'll bet seven or eight cases of beer, maybe a fifth of Scotch, on it. I would bet an equivalent amount that Mr. Chandler did not search out those places at first. No, the places searched him out. He did his job day after day, and those places struck some kind of chord every time he passed them; and finally after all those attempts to communicate with Mr. Chandler, the Sandhills tried it one more time and it worked. The next time, Mr. Chandler stopped at one of those places and thought some thoughts. Many attempts, none active, all passive, many failures, much happenstance and gamble, much fortuitous contact while doing other business, the business of simply staying alive, but those Chandler places finally came to Chandler, at some time in Chandler's life. Chandler places and Chandler times on the Chandler property. He now takes his time at his places. The conclusion may not be true, but it is what I am thinking as I open the gate between sections and look back to see Mr. Chandler, smiling and content and confident, behind the wheel of his pickup.

"The water tanks around those wells are full of snails," he says when I am back in the cab.

"Can we have a look?" The director smiles his own special

way, not necessarily a happy smile, nor unhappy, not really a smile at all, just that expression he has for a sense of impending biology. He probably also smiles that way when the Cleveland Indians make a double play. We jolt up to the tank; Chandler nudges the calves away with his pickup, though they never allow themselves to be touched. We are untold miles, at least twenty, from the nearest cattails.

The tank is full of *Stagnicola elodes*; so is the overflow pond. How did they get from here to there? These are Keystone Lake snails—I conclude they might as well be Keystone snails as Lewellen snails or simply snails from the next well tank down the road. There is something very satisfying about the conclusion that these particular snails came from Keystone. There is no way to prove such a thought, but Keystone Lake is as good a guess as any other *Stagnicola* place, so we'll go with it. The director and I stand in the mud of the overflow pond and turn over cottonwood leaves, sticks. Mr. Chandler joins us and we go through the standard biologist-at-the-mudhole routine. We pan up the microcrustacea, the ostracods, and we discuss what they might be. We poke through the pan at the aquatic insects. We note the whirligigs at one end. Biologist-at-the-mudhole routine: it's been done at least seventeen billion times; it's done every time the biologist walks up to a mudhole.

Mr. Chandler asks some questions. It is apparent that he did not really bring us out here to study snails or to pan up water from the overflow pond. He came here to sell us a couple of hundred thousand dollars' worth of Sandhills. The director and I rise to have a closer look at the tank itself. Mr. Chandler is still at the mudhole, poking through the water. There is more on the Chandler property than he realized. Mr. Chandler has grown children, although one is still in high school, and we talk about grown children a while. We also talk about things in the water.

"I wish I'd had the chance to study some of these things in school," he says, still looking at the ostracods.

"Come on down to the field station, any time," I say. The director is nodding, but looking in the tank. It is full of *Stagnicola elodes*. It is also full of some aquatic plant. Just as *Stagnicola* has found its time and place in the Chandler tanks, so has the plant. The plant is completely submerged; an *aquatic* plant, and it virtually fills the the tank. Any fool, any blithering idiot, can see the plant was not a yucca that had simply moved below the surface. The tank is also a plant time and plant place.

"You ought to see what comes out of these things when we clean them out," he says.

"Lots of stuff, huh?"

"Ever' kind of living thing in the mud down in the bottom of these tanks," says Mr. Chandler. "Really wish I'd had the opportunity to do like you fellas, study some of these things." I look at Mr. Chandler very carefully. There have been many times, very many times, in the last year when I wished oh so strongly for the opportunity to be something other than a university professor. This professor business is my place. There may have been a time when it was not my time, when had the right turn been taken, then I would be the rancher.

"You might find a bunch of professors ready to trade jobs with you, Mr. Chandler," I say.

"Not now'days, cattle business pretty hard to make ends meet." My mind is racing ahead now, way ahead, as the director palms the small *Stagnicola*. In my mind we have bought the Chandler property and there is a truck with a couple of dozen tanks on it parked just inside the gate.

"Where you want these, mister?" There and there and there and . . . there. We will see, we will, just how long it takes for *Stagnicola elodes* to return to a stock tank that has been cleaned out. We will see, we will, just how fast *S. elodes* can colonize a tiny bit of its place a hundred yards from the Chandler well, five hundred yards, a mile from the nearest *Stagnicola* population. In my mind racing ahead we will certainly test the ability of this thing to find its place in the world, we will determine *Stagnicola* time and *Stagnicola* place, and we will determine whether it happens every time, whether indeed *S. elodes* finds its place in the world every time. Still racing, my mind cancels the experiment; I know the answer. I may not actually know the answer, but I sense, I feel the answer. *Stagnicola elodes* will indeed find its place in the world every time. Pioneers. Any species that can find a Chandler stock tank can find its place anywhere in the world.

There is an uneasiness about this line of thinking, and it finally dawns on me. *Stagnicola* places are serene, calm, warm, protected. We think today that maybe there are not so many of these kinds of places, but we are very wrong. There are plenty of these places, and they are full of *Stagnicola*. The concrete pond between the city buidings. Moreover, the species has some device for finding its place in the world, for occupying its time in the history of the earth. Immediately I am envious of the snail, but I don't want it all. I don't want to be committed to a calm, serene, warm place and a calm, serene, warm time. There are times when I'd play *Stagnicola* however, times such as sitting on a log looking toward Lewellen.

Now I understand why I am so envious of the snail king. It has nothing to do with being in a serene place, a calm and warm place. It has everything to do with being in your own place, the place that is for you at the time that is for you. It has everything to do with a built-in device, a built-in inherited mechanism, for finding one's place in the world, regardless of how calm or serene, regardless of a maelstrom in the center of that place. I stand up, stretching at the end of the day, taking a last look toward Lewellen in the fading light, looking around, trying to decide whether to simply throw the empty beer can beneath the seat of my car or walk the hundred feet to a trash barrel. I decide on the car. *Stagnicola elodes* finds its place by being a pioneer. The string of eggs is the species' way of looking, exploring, grabbing every opportunity to hitch a ride to the next outpost, regardless. I will be a pioneer. I will try things I've never tried before. I will gamble some stuff, maybe I don't know just what yet; but if the snail king can gamble trillions of eggs, I also can gamble some things to find my place and time.

If I stopped here with *Stagnicola elodes*, the point of this tale would have been made, the point about finding one's place. The *Whaler* doesn't let me, however, for the *Boston Whaler* turns away from the south shore, away from the coves and into the busting waves, between the walleye fishermen, into the cold wind and spray of midlake. We are heading for the north shore; Otter

Creek and Sandy Creek. The mouths of these creeks are four miles away, but the *Whaler* slaps back at the waves as I slouch down to light a cigar. On the north shore is *Physa*.

We call it *Physa* in the same way we call another snail *Stagnicola*. Its shell is coiled in a different direction, and the coils are fat, becoming smaller with each turn, so that *Physa* has a very different look from *Stagnicola*. It is a good experience to study snail shells, to study them closely, especially with the aid of a book. The study of snail shells with a book must be but a fraction of the experience of studying them with the aid of a world expert, but there are not that many experts nowadays on snail shells, and the few there are generally don't come to Nebraska. It is easier to paint a picture of a marsh wren than to draw an accurate picture of a snail shell.

There have been many times when *snail* meant *Physa*. *Physa* is everything to everybody from all time back you can remember in almost every place you can remember. The animals simply made no sense at those times in history; they were under every rock in the dark, they were on every stem in the light. You collected a gallon jar of water from that pond and put it in your window, and within a few short hours there were *Physa* eggs glued solidly to the side of that jar. The sunlight would shine through the glass, through the eggs, and as the days passed the eggs would become embryos. You would get your hand lens and study the embryos for a while. Later they would all be baby snails spread out over the side of the jar. Them babies! Nobody studied *Physa*, but everybody knew *Physa*. To have said at the time you were studying the ecology and natural history of *Physa* would have been to invite looks of askance.

The distribution of *Physa* needs prior knowledge of the distribution of *Stagnicola elodes*. Study *Stagnicola elodes* in Keith County and you will then understand what it is to be a snail of the genus *Physa*. And that is why we are busting across the lake toward the north shore. There the mouths of spring-fed Sandhills streams empty into what used to be the North Platte River but is now Lake McConaughy. There are some mudflats, lots of dead trees, in places, and we will look for snails across the lake. I suck strongly on the cigar, inhaling a couple of times, and contemplate the difference between *Stagnicola* eggs and *Physa* eggs. The boat is slamming my spine and the wind is cold. There is one hell of a lot of difference in the egg masses. For one thing, I never saw a *Stagnicola* egg mass glued solidly to the back of another *Stagnicola*, but I can't say so much for *Physa* eggs.

The pictures taken at the mouth of Sandy Creek are among the most eerie of pictures taken that endless summer. These creeks like Sandy, Lonergan, Cedar, when crossed up on the highway, over bridges, are normal creeks. When approached from mid-lake into the mouth, they can be startling. There are acres of dead trees in the water, the sand deltas of these creeks slop out into the main lake, and the creek forms channels through the sand, through the dead trees. The channels meander. The water becomes very shallow. The hills rise slowly behind the trees, but normally there is no one at these places. They are not the clean and crystal places of Big Mac. They are instead the snail places, and the snail is *Physa*.

The pilot says that's all again, the Evinrude is gurgling, but this time we are far out from the trees. The water is shallow for a long long way out into the lake, and the boat sits there, getting smaller, as we wade into the mouth of Sandy Creek. The pictures taken that day show a couple of people standing almost waist-deep beside the big boat far from land; they show three or four people strung out, lifting knees high out of the water, against a backdrop of dead trees; they show a forest of leafless snags; they show a combination of things that is strange. The place is full of *Physa*. You knew it would be, you'd seen that from a couple of hundred yards out. Besides, you'd also been to those creeks in their upper reaches and found the animal.

Days later you would go to Lewellen, looking for *Stagnicola*, and find *Physa* along with, beneath the very tall cattails of the Lewellen marshes. You would wade up the mudflats of Lonergan and find *Physa* also, as you had found the animals in Keystone Marsh, at Ackley Valley, in fact about every place there had been water for you to step into knee-deep. Even the South Platte River. Yes, even the old South Platte, filled with *Physa*. Now that's an observation.

Pioneers

One asks sometimes, why it is that an animal species has been
"placed" on earth; what is the good of a mosquito, a housefly?
"Food for birds" is not a bad answer for some people. *Physa*
was placed on the earth for a very special reason and for a very
special group of people: it is here for the benefit of every human
who ever turned down an opportunity or found himself or herself
unable to cope, unable to extract some good out of a situation,
unable to break out of the warm and calm into the maelstrom.
Physa is here to offset *Stagnicola*. *Stagnicola* risks all to get serenity.
Physa is the libertine. Any animal that frequents the South Platte
River has a touch of libertine in the system.

A person can park on the north bank of the South Platte, turning off at the north end of the interstate exit viaduct required by
these towns, turning off into that flat parking spot. There is much
brush along the north bank, and places under the viaduct where
some high school kids have built a fire and spent the night. There
are also places where a skunk has been. The river is dry much
of its time and in most of its places, and you strike out toward
the channel somewhere in the middle of all that sand, willows
and yearling cottonwoods. There are many dried channels and
many chunks of driftwood. In the dried channels and beneath
the driftwood are *Physa* shells; they are there just as the *Stagnicola* shells will cover the banks of Keystone Lake some time this
year. Every dried channel has its shells; there are all sizes, the
South Platte treats young and adults alike.

There was this time down on the South Platte when twenty-one
people stood at the water's edge and talked of the complexities
of life. The complexities of life as a human are pretty simple
when compared to the complexities of life for a worm in the
intestine of a cliff swallow. Most humans are satisfied if they
find one place and one time, reasonably satisfied. The trematode
is obligated to a life that requires a number of places and times,
and furthermore the times and places must occur in a certain
sequence, a physical sequence and a temporal sequence. Small
wonder each worm has its built-in mechanisms for gambling,
small wonder that each worm matches its own set of mechanisms for gambling with an equally powerful set of mechanisms for

[33]

reducing the risk of that gamble. The swallows were nesting beneath the bridge south of Paxton, and I stood with the twenty beneath that bridge. The first requirement of a worm in the intestine of a cliff swallow is to get the eggs out. The second requirement is to find a snail.

"Put your finger on that event which is *the* interaction between swallow and environment that is required for the worm to complete the *first* step in its life cycle," I said. In writing now, the sentence sounds pretty formidable, very technical, perhaps out of some textbook. If you are not a biologist accustomed to reading this kind of thing, you might reread the question. Please reread your role statement, said the young committee woman. Standing in the South Platte River some student placed a finger on some bird droppings, cliff swallow droppings. The droppings were on some driftwood.

"Place in your hand the set of biochemical reactions, the highly coordinated set of biochemical reactions, required for that worm egg to hatch and to turn on the set of genes which code for proteins characteristic of the *next* phase of its life," I said. Now *that's* a tough question. *Now*, if you're not a biologist, reread the question. Standing there in the South Platte River some student picked up a *Physa*. The snail was on the same bit of driftwood as the droppings. It might have taken thirty seconds, but I doubt it, for the question to be answered. *Physa* is the highly coordinated set of chemical reactions required for the worm eggs to turn on the set of genes that code for proteins characteristic of the next phase of its life. The swallow above, the droppings and snail on the same bit of driftwood below.

You can read and reread in the books of this stuff about the lives of worms, the lives that are nothing more than role statements for worms and other animals alike. When extracted from the South Platte River and placed in the book, the worm story is one of high risk, high gamble, extreme waste for the benefit of a few survivors. The book without the South Platte is devoid of *Physa*, devoid of the catholic life of this simple snail, devoid of this creature's high-rolling, overcompetent, sure-gimme-the-stream-I'll-live-there approach to existence itself.

One must wonder now about the parasitic worm. The worm is

obviously dependent on the nutrients of the cliff swallow intestine, and as a larval worm on the nutrients of *Physa* liver. There is a higher level of parasitism here, however, for the worm is also dependent upon the *Physa* style. The inclination of this snail to explore, to pioneer, to survive in the wilderness of any roadside ditch or braided prairie river, to shrug off the South Platte as easily as it bathes in the luxury of Keystone Marsh, that inclination touches the life of this worm. All is revealed in the South Platte without the book. The *Physa* style has reduced the worm's gamble to a sure bet. The willingness of one species to gamble has touched the life of another and suddenly placed the other's obstacles into a different perspective. Any human standing in the South Platte who is able to see this and yet unable to let *Physa* touch his or her life is a human who will never be a pioneer. The snail spreads itself over Keith County, unabashedly, with total confidence, exploring, gambling, giving up a known place for an unknown. If the species had money, it is still doubtful that its style would change. Gambling with money, except in extreme cases, is a copout, it would say, and the returns are ephemeral. Besides, one has to have some money in order to explore the possibilities. A snail with no money steps into the unknown with the same thing every living creature on the earth has, with the same thing that every living creature that has ever lived on the earth has had: its life. In the single day you go from the Ackley Valley Ranch to any of a hundred little streams and ditches to the South Platte River to stinkholes to crystal streams to the mouth of Lonergan Creek, and you find *Physa* in every spot. Even a human must use *some* discretion, you think, but the life of *Physa* will not go from your mind. This is a good time to remind oneself of what Louis Agassiz had to say: "Study nature, not books!" I don't know if he actually added the exclamation point at the time.

It is fall now, and we have returned in September to the home of the snail king. Keystone Lake is virtually dry, a mudflat, and what is not dry is grown up with some aquatic plant in mats thick enough for a heron to walk on. The night is impossibly noisy, for the mudflat is covered with every bird that ever frequented a mudflat, or shallow drying fertile lake, on its way to some spot in the Southern Hemisphere. I walk to the small dock and throw a stone out over the mud; the birds arise in waves: ducks, geese, sandpipers, all out on the mudflats playing their roles, arising in waves to the motion and the stone. There go the eggs, the *Stagnicola* eggs; there go the eggs, the *Physa* eggs. There probably go some snails themselves, wedged between large toes, wedged up in some feathers. It is pioneer time for snails, a time to walk, a time to run, a time to go over the river and through the woods, a time to try it to see if it works, a time to get and keep a spot in some stock tank, a time to find some island in the sun of some Oklahoma mudhole. They'll each do it their own way, I think, in those warm and serene places there will be *Stagnicola* from Keystone Lake, pioneers who made it and who maybe will meet pioneers from another spot. In all those places, some serene and some turbulent, some clean and some dirty, some permanent and some temporary, will be *Physa* from Keystone Lake, pioneers who made it and who maybe will meet pioneers from some other clean or dirty spot.

It is still later fall now, and it is midnight in the capital city. I am in that calm and serene spot, that warm spot, trying my damnedest to sleep. In this country this time of year the geese go over at night and the wind blows. Tonight the geese are going over, I can hear them, their cries are cutting through the wind that sings off the bedroom corner of this house. There is some blowing dust, some leaves and twigs no doubt, in the city, and the dust and twigs rattle against the garage door. Funny, all I hear out there are snail eggs. That is not really a flock of geese, those are snail eggs getting from here to there. They might have loved Keystone, but they're leaving it, doing it their way. Tomorrow, I vow, I will pioneer somewhere, perhaps somewhere in the mind, but somewhere; I will take that time, simply, forcibly if necessary, take that time, to pioneer in some way.

This narrative has gotten out of hand. I retrace my activities of the last few weeks, and all I see are snail eggs. It is time to change the subject; it is time to talk of wrens.

4

Two Wrens

ONE IS SEEN and one is heard. The one that is seen is also heard, but the one that is heard is very rarely seen. Both are living bits of their places: the marsh wren is a bit of cattail head, broken, still clinging to a stick in the wind; the rock wren is a small bit of Brule with some very small lichens. They are both full of steam, as must every wren in the world be, and they couldn't have chosen two more different kinds of places to live.

It is difficult, once a person gets beyond the snails, to express Keystone Marsh in any terms other than those of the wrens. It is impossible not to study wrens, the long-billed marsh wrens, in Keystone. There are times when a person "goes after" a species, with equipment and intent, in order to understand the manner in which that species makes its way in the world. There are times when a person goes after a species with the idea that maybe, just maybe, there will be something in the life of that creature that will add perspective to the life of the person. These times are laced with other times, however, when the unexpected species comes forth, stands out and up, gets the attention, makes the point.

A pair of fishermen drift down the canal. A canal has been dredged along the south side of Keystone Marsh, a canal to keep the water flowing, and there is a large sand dike, the dredgings, on the marsh side of this canal. Small strands of cattails have made it over the dike, and every stand has its pair of wrens. A wren comes to the top as the boat approaches. The insect buzz (the word *scold* has never been applied more appropriately) is directed at the fishermen, who appear not to notice, and the task is picked up by other wrens down the canal as the boat nears that place just past the bridge where the canal empties into the lake. The wrens have done their stunt for the day, they've gotten the fishermen down the canal! I am not able to understand how a fisherman can drift down that canal once and not become an ornithologist, a studier of wrens. I cannot understand what is going on in that boat, what kind of conversation, what kind of equipment preparations, that could override the marsh wren. A fisherman by all rights should simply throw the tackle overboard, beach the boat, and wade in after the wrens. That's about the effect *Telmatodytes palustris* has on me and all my friends, regardless of the reasons for entering the cattails. One can collect snails without thinking, but one cannot collect snails without listening to the wrens. Before long the listening pace quickens and the collecting pace slackens. Before much longer, the snail picking stops altogether while the back is straightened, and the knuckle sucked. The buzzing comes closer very rapidly. The wren is there, a few feet from your head, then all is quiet. Back to the snails, but the wren is there also, this time silent, slipping along beneath the cattails. Eye to eye with a marsh wren beneath the cattails, then the bird is gone and the person is back to the snails until, not long after, the ritual begins again. The marsh is alive and well.

There was a person that summer, a cheerleader, a Girl Scout, who always liked to do that kind of stuff.

"Sure, I'll do it, I like to do that kind of stuff!" she used to say. Such statements are statements of responsibility, they put the volunteer into any kind of position, usually one of responsibility. It is easy to foist responsibility off on a volunteer, and subsequently it is just as easy to forget that the volunteer is shouldering the responsibility. I have this impression, this feeling, this almost-dream, that there was a time, a prehistoric time when mastodons roamed the plains of Keith County, when a member of the Nebraska Game and Parks Commission drove a state car out to this place and asked for volunteers.

[38]

"We need a volunteer," he said, surveying the animals, "we need someone to take the responsibility of letting us know the cattail marsh is alive and well." He was a very serious person. People with state cars have a job to do.

"Sure, I'll do it," said the marsh wren, "I like to do that kind of stuff!" I also have this feeling that Game and Parks has forgotten the wren, has after these thousands of years come to take the thing for granted. The bird is still there, boys down at G and P, still doing its job, volunteer work. Maybe with a little support it could even become a tourist attraction. On the other hand, the Game and Parks commissioner went to the doctor today, and the nurse put her hand on his wrist. Feeling the pulse, she declared the commissioner alive and well. No pulse ever became a tourist attraction, no matter how long it has carried the responsibility of being a monitoring device.

People enter a marsh for various reasons. The act of entering a marsh is highly recommended, and as a teacher I find it comes close to being the very act that breaks the shell of inhibition, that cracks the unwillingness to participate, in a student. If the act of breaking and entering a cattail marsh does not do the trick, or if seining alone does not, then the combination is surefire. Any student who fails to participate after breaking and entering a cattail marsh, especially if the act comes on the heels of a good seining experience, has serious problems. Seining is an activity that really is not, in its finest form, readily available to the public.

Anyone can enter a cattail marsh, however, usually without permission, since there are plenty of discarded ones on public land. Just wade in. Just walk up and wade in. It's that easy. Just enter the marsh. Oh, most assuredly there is mud, often vile, beneath the water, and the cattails are difficult to walk through. You flounder, sometimes even fall right down in the stuff. It's the first step, that's the big one. Ready now? Your shoes and pants are dry? Stand on the sand and go. Lift the foot and simply place it, shoe and all, in the water, then push the shoe down into the water and the mud below; keep pushing, knee-deep? Okay. Now try the other one; see how easily it goes in right beside the first? That smell? Oh, that's the marsh. Yes, your tennis shoes

[39]

and jeans *are* going to smell just like that for the rest of their useful lives. That buzzing sound? That's the marsh wren. Now you've placed your muddy feet on the marsh, the wren-pulse is telling you the marsh is alive and well. Yes, every time you touch the marsh you will feel the wren-pulse. At least as long as the marsh is alive and well.

The wren is everywhere. Across the lake now in a canoe, hunting muskrats on the north shore. The muskrat stunt is tried, slipping very quietly along the cattails, not extensive but certainly all along the shore for almost a mile, looking at those places where muskrat might be sitting, nibbling, twitching its nose, waiting for a bullet. Then there is the buzz. A substantial attempt is made to ignore it. After all, this is a sneaky expedition. Not a chance. The buzz picks up as the canoe drifts; in the lee of the north shore, a canoe drifts and glides with but a touch of the paddle, for the north shore of Keystone Lake is also a dike. It is very hot, late in the afternoon, although Keystone Lake has no mosquitoes, none to speak of, only the wrens. The hunters are distracted now, the rattling buzz becomes more intense, but the muskrat is seen too late. There was a chance for a shot, had they been alert, but now there is only the most fleeting of backs and tails. The hunters will look well ahead this time, will find those runaways well before the canoe drifts by. It doesn't work. There is still this distraction, and not only that: a duck, flightless or injured, flops out of the cattails and is gone. Ripples wash down the shore for a hundred yards or so and there are a couple of responses from the wren-pulse.

Later the hunters will try the same thing with yellow-headed blackbirds. They will try the same thing with black terns. They will try the same things with mice, with small traps set at the edge of the marsh. Sometimes these things will work, sometimes not. The activities will be carefully watched, however, carefully recorded and upon occasion, I feel, passed on to the next generation of wrens, the keepers of the keys to Keystone Marsh. In the eons since the mastodons roamed, Game and Parks may have forgotten the wrens, taking for granted the work was being done. In those same eons of volunteer work the wren has extended its own responsibility to include a watch over the marsh. First the pulse, then the eyes and ears of the Keystone cattails; the roles are very different. A creature, given an onerous task, completes it willingly and easily, and in the course of time comes to view the task not a task but a role, a place in the network of the world's living society, not simply some commissioner's idea of a way to keep busy, a way to ensure every tiny bit of the state's business is done. Sure, I'll do that kind of stuff, says the wren, I like to do that kind of stuff.

People enter the marsh now, people who have broken and entered before, young people, people who for the first time in their lives have a marsh at their doorstep. They walk across the heavy wooden bridge and turn west along the dredging dike for the spot. They linger along the way, picking at rocks, picking at empty snail shells, picking at killdeers, looking for killdeer nests, sucking their knuckles to tease the blackbirds, skipping rocks in an occasional spontaneous contest, sending someone back to get the gallon jars forgotten in the morning's preparations. They are not anxious, they are perfectly relaxed today; not like the first day, that first day when some felt a need to prove themselves, when some were very apprehensive about their own abilities to deal with any situation as complicated as the cattails.

On the other side of the dike, in a small pocket of cattails, the wren starts. They learned the wren the first day, they learned to suck up a wren on their knuckles, and it was almost the first thing they learned about the marsh. They learned it from the old man who learned it from another old man, the latter the best in the knuckle-sucking business. A sucked knuckle, if done properly, sounds like a baby bird in real trouble. When done by the best in the business, it sounds like a bird in so much trouble that there is no way in all living hell that poor creature is going to be saved, a fledgling blackbird being eaten in slow motion by a raccoon but screaming the screams of a whole marsh full of baby blackbirds in fast time. Back in those Oklahoma woods such a knuckle suck sent shivers down the spine of even the humans in the group. Come on now, adults, get here quickly, come on up, make this man stop this infernal sound. Out in the marsh

[41]

even the most simple knuckle suck brings up the keeper of the keys, instantly. They all learned that the first day; now they all do it. They do it to the wren on the other side of the dike. They stop down at the entering place. A few gather their gear for a stomp through the cattails, but a few others always call up the wrens.

I stand in the morning sun. Today we are going after land snails in the cattails. The species is *Oxyloma retusa*, and although they live in the marsh they are land snails. They never get wet, to my knowledge, on purpose. We have never found an *Oxyloma retusa* with muddy feet, just as we have never found a *Physa* with clean feet. I wonder today, standing in the heating sand waiting for the group, what would happen if no marsh wren answered the suckers. I have this feeling that it would be first day all over again. The group has come to depend on the keeper of the keys. All it takes is once. One time, that is all. Step to the marsh, call the keeper, the keeper answers, and it is all right to go into the cattails. The feeling is gone now, and a sureness is there. I am totally confident that without this ritual, that if no wren answered, the young people would not go into the marsh. The animal that liked to do that kind of stuff volunteered back in the time of the mastodons, the animal served well and became the keeper of the keys, the eyes and ears, and now he gives permission. The humans require the permission; they have come to depend on the keeper to do his job. This morning there are wrens and they answer. Permission is given, and into the cattails we go after more snails. This is still a tale of wrens, however, for the life of *Oxyloma retusa* is intimately tied to the life of the wren; or so we say, since we refuse to kill enough wrens to find out for sure. Some questions are best left unanswered.

Below the cattails it is silent, the water is calm regardless of a gale that clatters the tops. The half-submerged stalks of last year's growth form an uncertain mat for human feet, and there are places where it seems a deer may have lain; or run. The water is warm; in small and almost dry pockets there may be some mosquito larvae, carrying protozoa on their backs. There is an unbelievable number of insects. There is an almost equal variety of spiders. There is seemingly a spider and a spider type for every insect and insect type, for there are many many spiders indeed below the cattails. There are caterpillars in the cattails. There are real caterpillars and fake caterpillars. The wren consumes all these things. The wren is the dominant form in this marsh, and one could study, as did Kale in Georgia, the flow of energy through the marsh into the wren. Energy flows from sun through plants through a layer or two of insects and spiders and into the wren. The wrens have much energy! Wait a moment, the fake caterpillar is also a part of this food chain, the fake caterpillar has plugged into the series of events. The fake caterpillar lives in the tentacle of *Oxyloma retusa*. The fake caterpillar is really the larva of an intestinal parasitic worm.

The worm is a member of the trematode genus *Leucochloridium* and lives as a larval form in land snails, but as an adult in the digestive tract, or associated organs, of a bird. The snails are difficult to see, normally. Often an effective way to collect *Oxyloma* is simply to stomp down the cattails, submerged, then let the affronted snails crawl up your leg. An infected snail is pretty easy to see, however, for the tentacle is a brightly colored fake caterpillar. The larval worm has worked its way up into the inside of the snail's tentacle, it has synthesized or stolen some pigment that it has laid down in green and brown bands, and it pulsates, throbs, in and out of the tentacle. The human stands there looking at the infected snail beneath the cattails. It is the tentacle that caught the eye, and most assuredly the tentacle that catches a human eye will be sure to catch a wren eye. The slender but vocal bill that snaps up *this* caterpillar will send to the digestive tract maybe a hundred, maybe a thousand, parasitic worms. Let's see, first a volunteer to act as the pulse of Keystone Marsh, simply to let us know the marsh is alive and well. Next as the years go by the eyes and ears, the dominant form in this discarded portion of the ecosystem. Now the giver of permission, the keeper of the keys. An onerous task well done, well done for the benefit of society, and now because of it a position of responsibility, stability, a role. Now comes the bitter with the sweet; the rip-off. I reach for the infected snail but the wren beats me to it.

"Don't." I say; but it is too late. The fluff has shaken the fake caterpillar and consumed it in one gulp. This cannot be the first time it has happened, nor can this event really have much effect on the dominance of this marsh by the wren population.

It is a football morning, silent, cold, crystal, and the lingering call of a magpie filters down from the next canyon, or maybe the next one, or maybe even the one after that; no one knows how far the magpie call will filter on a morning like this. The station is deserted, and I stand on the back patio, with coffee, looking over Keystone Lake. Way down the hill, fumbling around in the brush, is a cream-colored station wagon with a red picture of the state of Nebraska on the front doors. Surveyors. I walk toward the boat house and there are targets on the sand, aerial survey targets, large wooden sticks with radiating red plastic strips tacked to the sand. The surveyors are here to take the marsh. There is a hydroelectric plant scheduled for the spillway of Kingsley Dam, and the diversion dam at the eastern end of Keystone Lake will be raised three feet. The water level of Keystone itself will be raised three feet. It was not my feeling during the endless summer that this was actually going to happen, regardless of the local talk. Now the men are here for the marsh.

"What's this trash?" I say. There are three of them and they are burly. The station wagon is filled with survey equipment, poles, transits, what else, and the tires are worn. The Nebraska emblem is also worn. This is a field team. Engineers. The men look at one another, then back toward me. One goes back to work, the other two have very blank expressions.

"What's this trash?" I repeat.

"Ain't trash, mister; aerial survey targets."

"Have you asked the wrens?"

"Huh?"

"Have you asked the wrens, you dumb shits, have you asked the wrens' permission to take the marsh?"

"Who you callin' a 'dumb shit'?"

The conversation really does not take place anywhere but in my mind standing on the patio, watching the engineers fumble through the brush far down the hill below. Of course they have not asked the wrens' permission to take the marsh. If questioned, Game and Parks would formally state that the marsh wren is responsible only for telling whether the marsh is alive and well, not for permission to take the marsh. I turn back inside, wondering if there is in fact anyone anywhere whose permission must be asked before the marsh can be taken. No wonder the bird snatched that infected snail tentacle out of my grasp. A gut full of worms, now that I think about it, stoking the fire, is not much of a rip-off after all.

Salpinctes obsoletus is a very plain name for a bundle of fire known as the rock wren. It is heard, up on the bluffs, up in the rocks, but it is seen only by those who climb the bluffs regularly, and then it is seen only irregularly. Except today. For some stupid reason it had done the thing that wrens all over the world must do—fly into a small spot and get stuck, trapped, when the small spot turns out to be a cavern. Tom Sawyer and Huck Finn. Today it flew into the men's wash house and was caught, caught and placed in a small hardware cloth cage. The cage was a bit of totally ingenious prior preparation by a field man's field man, for on one end it had a door. The door was simply two strips of rubber inner tube, nailed so that a hand with a wren could easily be slipped between them, but a wren without a hand could never find the entrance from within. There was a worm man guarding the cage. To a worm man a rock wren is a dish full of intestinal contents. To a wren man a rock wren is a model to be painted. I bargained for its life, or at least for its temporary survival.

"I will put the damn thing back in the cage," I said, "I will put the damn thing right back in the cage and put the cage right back in this very spot."

He looked very suspicious and stroked his beard.

"I do admit this very publicly. This is your bird, your bird that you trapped while cleaning the wash house, I admit this freely, and when I'm through with the painting I'll return your damn bird right here to this very spot!"

He was not satisfied, but the bird was taken away, back to my cabin, away from the din of business and curiosity. I had vowed not to release it. Not to release it, to relinquish it to its "rightful

[45]

Long-Billed Marsh Wren

". . . the marsh wren is a bit of cattail head, broken, still clinging to a stick in the wind; . . ."

owner," meant sure death for the bird. None of this bothered me at the time, for I had spent a few hours in bird intestines myself. It did bother me to be accused of plotting to release the bird.

As you will read later, I am a painter of birds, although through no fault of my own. There are things you do because other people you know do them, and painting birds is one of these things. I am not a very good bird painter, and you will not learn much about painting birds from the chapter of that title. All of the value in painting birds falls to the painter rather than the painting. The pencil sketch took about thirty minutes. The pattern on the back of a rock wren is exceedingly complex and I simply had no idea of how to begin converting watercolors into that pattern. So I started with the eye.

I normally start with the eye, and the eye is normally the easiest. But I don't start with the eye because it's the easiest, but because if the eye is not the right shape, the pupil placed correctly, then the bird simply does not look alive. Incidently, the foot is the hardest, anatomically the least understood, anatomically the most difficult to manipulate and represent on paper. The best way to ruin a good eye is to put a bad foot underneath it. Models are normally very cooperative with the eye, and this one was no exception. I held the bit of fire by the feet, between the fingers, so that it faced toward my left. Often a bird will calm down in this position, calm down somewhat. Not the wren— not calm, but then not jittery or uncontrollable either. Just a ball of fire. A twit. It glared; it was baleful, insulting, affronted, darting and full of the emotional stuff, with its tarometatarsi wedged firmly between my fingers. I apologized. I almost asked its permission to drop some water colors on my own paper. What is it about a wren that gives one the feeling that permission should at least be asked?

Something had come down out of the hills with the bird, something wild, something heard off in the toughest part of the canyon but never seen. It communicated; it told me of the hard times up on the outcrops, and I believed the story, but I believed the story knowing full well it was a lie. I was being put on; life on the outcrops was a bowl of cherries. There was plenty to eat up there, and any creature with the flitty equipment of a wren is made for life on the outcrops. This animal came from my backyard, but from a spot in that backyard where I really could not go. Something told me this one knew nothing of the marsh wren. If the rock wren had known anything of the marsh wren, I think the former would have had nothing but pure disdain for the latter. It's all in the test, I was told that day; live in a place where you are not tested, and you are living in a place of inferior quality. Come to the outcrops. Come to the outcrops and you will be tested. Come to the men's washroom, big shot, and you will also be tested, I replied; then I came to the feet.

Models may be very cooperative with the eye, but they are very uncooperative with feet, and if allowed to place the feet naturally will normally fly away. This one's feet were very gnarled, almost as if hopping around on the Brule outcrops was hard manual labor. I worked hard on the foot, realizing that any creature that comes with gnarled feet is a creature that does have, somewhere in its life, some contact with what is physically tough and hard. The picture was finished now, as best I could manage. The pattern on the back was hopeless. One might have to paint that rock wren fifteen or twenty times to get the feel of how to make the pattern, much less be able to make it. I was glad to finish the picture and rid myself of the hill thing that demeaned my every movement. I see now, after writing this, that I know and feel absolutely nothing about the rock wren aside from what was learned that day painting. I also have this feeling the rock wren prefers it that way. I returned the thing to its cage, not really caring whether it was converted into tapeworms. I returned the cage to the laboratory and covered it with a cloth. I went to dinner and played volleyball afterwards. The wren was forgotten; it belonged to the worm man.

"What'd you do with the wren?" he said in the morning.

"I put it back in the cage."

"No, you didn't."

"Yes, I did put the wren back in the cage and put the cage back in the lab."

"It's not there now."

"I don't know where your bird is; I put it back in the cage."

"You're sure?"

"I'm sure; I told you I would put the bird back in the cage. You're sure it's not hiding in the cage?"

"How in the hell could a bird hide in a cage like that!" (I agreed.)

We checked the lab. Wrens that are loose in the room have a very distinct way of flying about. The animal was gone.

"Could Cindy have let it out?"

"I have no idea." Cindy is my daughter; she will explode when she reads someone thinks she might have done something on the sly like that.

"Sorry, Rich, the wren is gone. I did not let it go."

He grumped for a while.

Later that morning, relaxed now, I climbed the hill in back of our cabin and picked a rock to go under the painted wren. The rock cooperated beautifully, and even seemed to enjoy having its picture made. I still have the rock. It is a wild rock, from the hills on the south of Keystone Lake, but it does all the things a pet rock from Bloomingdale's ever did or ever will do, and it has adapted well to human company.

The wren must be back on the canyon wall now, and I sense an overwhelming lack of knowledge, a lack of familiarity, a lack of introduction, to this creature that entered my life so suddenly and under such quick circumstances. It was here and gone. I go then to the library to find out about the rock wren and am shocked. After reading even the most elementary writings of the rock wren I am shocked at society's ignorance of this bird. Why has the rock wren not been discussed, over and over, in every textbook that has ever been written? Is it because no one has taken the time to read what is known about the thing? Dawson tells me the rock wren makes its nest in the rocks, in the nooks and crannies; but I knew that. Dawson also tells me the rock wren makes its nest *of* rocks, and that startles me. *Of* rocks? They are evidently flat little pebbles, these rocks, selected with the utmost care by the frail bill, and placed carefully in a bed. A bed of rocks? The wren actually makes a bed of carefully

Rock Wren

"Something had come down out of the hills with the bird, something wild, something heard off in the toughest part of the canyon but never seen."

selected rocks? Is there a creature like the marsh wren that becomes so at one with its environment that it becomes a busted cattail? The marsh is full of cattails; they are there by the hundreds, the thousands. The marsh wren makes many nests. Is the animal that has become a busted cattail trying to emulate the marsh and become many busted cattails? I accept the fact that it might be. Dawson tells me the rock wren does the same, makes many nests. Is the animal that has become a rock trying to emulate the Brule outcrops, the very bluffs where it lives? I also accept the fact that it might.

Dawson tells me the rock pile of the wren's nest makes a "pleasant tinkling sound" when the wren leaves. I know now what must be done the next endless summer: I must hear the sound of the wren leaving its nest. Oh boy; oh boy, oh boy! This defiant creature with the gnarled feet makes a bed of rocks that tinkles like a wind chime so many times a day for so many times a summer? I know the thoughts that will come when I hear the wren rocks: tens of thousands of years ago people lived in nooks and crannies in the bluffs, even in the bluffs of Keith County, and as they lived there they modified the crannies, making them art museums. I feel now they must have made them also concert halls, those caves in France where they drew the pictures of their bison and rhinos. They must have made some music, must have noted the sound of the wind in the bluffs and arranged something to make a sound, not an imitation, but a sound with the equivalent subtlety. I see Neanderthal now, entering his cave, stepping on the bed of small stones, noting the pleasant sound of crunch and tinkle, assuring himself that after all he was home. Yes, I will climb the bluffs and listen to the wren nest.

5

Swallows

THE PERSON to give me the most recent flak about my seventy-dollar martin house was my ten-year-old daughter, but that was before the martins came. She had that look on her face, that forced smile, the blistering eye, that said, "This is mental rowdiness." She had asked how swallows fight sparrows. Before that, even the day I brought the martin house home, the neighbor had given me flak.

"I see you bought a sparrow house!" He laughed. His family, getting out of the large station wagon, smiled. I shudder to think what it would have been like if he, like my wife, knew what the house cost. Nearly $70. I have spent a great deal more than $70 and not received nearly the pleasure I received the first time a martin actually flew by closely and took a good long look at my martin house. I can think of a $183 brake job on our car, purchased while on a family vacation in New Orleans, that did not give me nearly the pleasure of my martinless martin house. Even the act of putting up the house gave me more pleasure than the brake job. I turned the house so that I could see the apartment openings from my place at the kitchen bar. This was all in March. I nearly had a wreck and heart seizure the day the martins returned to town.

It was about two weeks after their return that a single martin actually came by to look closely at my house, and with that one event I felt I had received my money's worth. It was another two weeks before the next significant event happened. A large male martin flew up to the house and challenged a sparrow for the physical possession of the property. It was the last time I saw that particular martin. The house had now returned several times the original investment.

The depths of depression, of course, came later, when I realized that martins were very busy with their business of establishing homes, but in every other house in the neighborhood except mine. Two families of sparrows were now firmly in control of the property. I have always been an opponent of sparrow-nest destruction, feeling that prevention if possible was the best alternative. Prevention in this case meant almost fifteen hours a day of standing below the house and frightening the animals away. A sparrow is a very persistent creature, also subtle, businesslike, and skilled. I removed the entire sparrow nest and buried it; a little tolerance on the part of the sparrows would have saved their family. My mind had been at work seriously the day before, analyzing the problem of why there were no martins in or on or even around my house.

Two martins had actually landed on my house, early in the morning. The sparrow was upset, very upset, but the martins were vocalizing with a wide range of whirs and grates. It was a few minutes before the final burst of sparrow aggression was released. The martins had actually entered some apartments and returned to the top of the house to continue their discussion. At this point the male sparrow had had enough, and responded with angry chirps, very aggressive bluffing and defense of the territory, repeatedly flying at my seventy-dollar martins. The martins were driven off. I don't know if they are the same birds now in residence, now that the sparrows have been subdued, but on that day they were driven off. This was the day the question came about the way birds fight.

I like to think my solution was actually a solution, that my correction of the problem was the event that actually brought the martins. Swallows fly fast, in driving long spurts and rapid glides; they swoop over fields. Although they may nest in a hole, a swallow on the wing suggests an animal capable of much

[53]

claustrophobia. I sat at dinner, watching the sparrow drag up dried weeds that had so recently come from one rose bed. It occurred to me in a flash of brilliance that the apartment doors pointed between our home and that of the neighbors, a narrow channel through which no swallows ever flew. I turned the house ninety degrees so that the apartment doors now opened to the length of our yard, and the view from the house was of a block of back yards. The martins came instantly. Three females. They would not be dislodged by the sparrows. They are still there now. They fuss a little when the kids play prison ball or when I pick weeds from the garden directly below their house, but that too will pass soon. They are beautiful and I watch them continuously from my deck door, with binoculars. It must be all of thirty feet from the deck door to their house, but I still use ten-power binoculars. It is conceivable that in the near future I will buy a telescope just to watch the purple martins thirty feet from my deck door.

There are still sparrows, but the fighting has stopped; evidently both sides are reconciled, and it will be several years before the entire house is taken up by martins. There has been no physical combat, only angry vocalizations and bluffing. One gets the impression the sparrow might actually engage in contact fighting, but the martins are very careful about what they actually encounter physically. Any bird that purposefully takes seven wing beats to get from the edge of the roof to the top of the roof of a purple martin house is a bird that likes to use its body. I had answered my daughter that birds generally fought by acting tough, and that the one that acted the toughest usually came out the winner. Even to a ten-year-old, this explanation was far enough outside the human concept of fighting to elicit a smile. Humans bludgeon one another, even on television, and shoot one another, even on television, and SWAT one another, even on prime-time evening television. Humans say that to be a human is to have a mind that surpasses in its creative and learning capacity that of all other kinds of creatures. The mind is what makes a human a human. To be a swallow, though, is to have a body that surpasses in grace and coordination that of all other kinds of creatures. The body is what makes a swallow a swallow.

Humans and swallows fight with the thing that they are not, bringing into the battle the thing that they are only if forced beyond the point of choice. It's all right, biologically correct, to show the bludgeoning and SWATing of human bodies on commercial television. The protests would come if the commercial media depicted humans actually using their minds, in a convincing and real manner, to precipitate more injury on fellow man or competitor than they could using their bodies alone. The swallow-sparrow conflict is mental, and the outcome dependent upon participant interpretation of the bodily activity. I cannot believe a purple martin would actually touch an English sparrow with its precious body unless truly forced to. This simplistic idea is confirmed strongly and completely, if one only examines a barn swallow nest under the cement culvert on the Keystone to Paxton road.

The Keystone-Paxton road is made of sand and cattle guards, and runs generally parallel to the North Platte River, having crossed that river at the Paxton bridge north of town. Paxton is down the asphalt highway a few miles from the roadside ditch where the spadefoot toads breed. The town is world-famous for Ole's Bar, and Ole's is famous for its animals. Just inside the door is a polar bear killing a seal. Further inside are some equally dramatic prey-predator relationships. Covering the walls are mounted specimens of every conceivable kind of potential "game" animal, ranging from a king vulture to a marine iguana, with the normal series of wart hogs and antelope between. I found Ole's very educational; there are not many zoos or museums where one can drink beer in the middle of the day and hold class to the strains of Merle Haggard or Hank Williams, Jr. Cliff swallows are starting a new colony beneath the bridge north of Paxton. The road turns left across a cattle guard and heads out through termite country to Keystone.

Keystone is on the north shore of the North Platte River and consists of a special small white church and a laundry–grocery store–filling station, a ball diamond, a gravel fork in the road, and a few frame houses with no one around. It was near Keystone that Killer was captured, running across the water in a roadside ditch beneath some small stray willows. A brave and healthy plains lad pursued Killer with a mayonnaise jar and deftly scooped her up just before she reached the shelter of the dense brush. Ensconced in a gallon jar labeled "home" and "visitors," she promptly dispatched all comers, including a large centipede from under a local rock. Killer was never identified as to species, and I am not sure our spider literature is complete enough to accomplish that even if we had tried. Not far from Killer's ditch is *Hirundo*'s culvert, a very special kind of culvert, transmitting a spring-fed creek beneath the Keystone-Paxton road, and containing a concrete slide, which is great fun to play on, and barn swallow nests. The nests are snug up against the ceiling of the culvert. A swallow appears always to cover its back, if not with a seventy-dollar apartment or a mud flask of its own making, then with the ceiling of a culvert. The barn swallow nests are mud and dried grass and are stuck to the culvert wall just under the ceiling. There may be many nests in a culvert, but one always gets the impression the gathering is fortuitous, and that the animals are not really dependent upon one another's company, or at least not directly dependent in the manner of purple martins and cliff swallows.

Barn swallows are incredibly fast. They are incredibly streamlined, especially in the air. They are incredibly numerous, and abound in and around culverts on every mile of every road in the United States. And they are incredibly widespread; a visitor to Turkistan, on a once-in-a-lifetime jaunt, spending the insurance money for a ticket from Keystone or Paxton to the rest of the world, will see a barn swallow buzzing the road over the first culvert outside the airport. In the hand they weigh nothing. They are all wings and tail. In the hand they are proud, a temporarily grounded thing of the air. They look at you totally convinced they are only temporarily grounded. The look is not only convinced but also very convincing, and the human usually concludes that indeed, after all, this inconvenience is only temporary, and the barn swallow is released. Free, they act elite. Released, they make a mockery of the word *flight* as we normally apply it to the birds. One thinks of birds in the same thought as one thinks of flight; however, few birds fly as do the swallows and few swallows fly as do the barn swallows. To know a barn swallow, to handle the bird, to paint its picture, is to reassess one's impressions of what it means to be a bird. A bird, your average run-of-the-mill bird, simply does not know what it is to fly any more than a human being does, compared to a barn swallow.

The barn swallow's final insult to the average landlocked bird was immediately apparent when we examined those nests under the culvert under the Keystone-Paxton road. *Hirundo rustica* had lined its nest with feathers of the great horned owl. Having handled a barn swallow, my first thought was that the *birds* had concluded that no creature that spends most of its time sleeping has any need for feathers, and had simply plucked them, while flying, from the breast or flanks of the owl. A better explanation is that they probably found them. A fairly obvious conclusion is that swallows believe swallow feathers belong on swallows, and that there are plenty of lesser birds to suppy parts of *their* airfoils to line a swallow's nest. It is difficult not to agree. After all, no Olympic-class hurdler digs a garden in her track shoes.

There are things that grow on human accoutrements and structures: molds, mosses, fungi, mildews, and cliff swallows. One must sense that these growths were present on earth before man, and certainly before man's structures, and in those times before man's structures simply grew on wild things and in wild places. With the advent of civilization, these growths have found fertile new environments to exploit, and the marketplace is filled with chemicals to prevent the civilization of these growths; except for the cliff swallow—there is no chemical on the average grocery store shelf to prevent the growth of cliff swallows. On a culture plate a bacterial colony or fungus will fill the available space and finally assume a characteristic form before, as a colony, reproducing or producing the characteristic reproductive forms. Certainly in the case of a fungus a set of defined developmental events ending with the production of a fruiting body will occur, normally in a genetically-determined sequence. One should interpret a cliff swallow in no less a manner: their colonies are initiated by a few pairs; the philosophical equivalent of fungal spores,

Barn Swallow

"... *your average run-of-the-mill bird, simply does not know what it is to fly any more than a human being, compared to a barn swallow.*"

[58]

they get a hold on some unused corner of a human structure before the human notices them, and they proceed to flow through the available space as rapidly as the years and multiple broods will allow.

Mr. and Mrs. Retired Couple in a small Winnebago leave I-80 at the Ogallala exit. She wipes her brow in the July afternoon and he, driving, squints through the sun at the monstrous plastic football player atop a local restaurant. Vaguely she perceives the South Platte River, now only a trickle, while she glances at the Ramada Inn on the banks, wishing down deep that they could check in rather than spend another night in the camper. The South Platte at this point elicits no positive response, for in the best of times only a few call the South Platte, or any other prairie river like it, beautiful. It is all sand, a few puddles, a small, very small, channel, pilings and driftwood, yearling weeds that have taken a hold since the spring floods, and above all, green algal mats.

Mrs. Couple loves birds. Back home she has several wren houses and a large feeder where blue jays and chickadees come during the winter. She has spent considerable time figuring out how to keep the squirrels off the feeder, since the male cardinal doesn't show up so regularly when the squirrels have eaten all the seeds. She tried a hummingbird feeder one year, and it didn't work. But today it's July on the Ogallala exit ramp, and the swarm of cliff swallows that dart under the South Platte exit bridge and swirl like blowing leaves around the camper doesn't register on her. Down on the dried riverbed a couple of boys in cutoffs and long T-shirts are throwing rocks up toward the road. They should be home, she thinks, and wonders if their mother knows they are down on this river, where there is probably quicksand, throwing rocks at cars up on the bridge. A rock hits a cliff swallow nest and the nest, totally fragile and made of sand held together with wishes, disintegrates into powder and a couple of bigger chunks. Unfeathered nestlings fall to the July South Platte sand. Those on their backs tread air feebly, those on their stomachs try to lift their heads. The boys bend over the birds briefly, then search for some more rocks. Overhead, under,

around, and through the bridge the adults are swirling. The sand
is littered with egg shells and some nestlings, some alive and
some dried. The raccoons will clean up the live ones tonight.
Over the next several days the ants will clean up the dead raccoons up on the interstate. The decomposed dead ants will be
incorporated into the soil of Keith County. The soil of Keith
County, next year, will be incorporated into cliff swallow nests.
In the meantime, the bridge colony copes with young boys by
flying and twittering, in large groups. It is significant that there
is no bird that dives, pecking, at the young men. Tired of their
sport, they wade ankle-deep into the channel and up the river a
mile.

Mr. and Mrs. Couple drive on to Big Mac, where they are
afraid to drive their camper onto the sand, so elect to take the
gravel road below and spend tonight on Lake Ogallala, the bor-
row pit for Kingsley Dam. It is midge night at Lake Ogallala, an
event the Couples did not know was scheduled until it was too
late to find another pace for the Winnebago. Just before dusk,
however, Mrs. Couple, relaxed now in the cooling evening, does
notice the swarm of cliff swallows over Lake Ogallala. It is with
identifiable pleasure that she notes these birds, for Mrs. Couple
likes—no loves—birds. She smiles, stretches, breathes deeply,
and the first of tonight's quadrillion chironomid midges is sucked
into her throat. While she coughs and convulses, it never dawns
on her that she has just tasted cliff swallow food!

There is no such thing as a single cliff swallow. The individual
is subordinate to the colony, and it is the colony that grows,
occupies the available space, and reproduces, amoebalike, new
colonies. The birds are not difficult to catch in an insect net. Late
at night with a flashlight and net, one can get ten or fifteen in
a single sweep along the mouths of the nests. In a wire cage the
ten cliff swallows orient their bodies in parallel, their eyes follow
one in unison, their heads turn synchronized, they are docile,
they do not fight the cage nor do they attempt to systematically
dismantle it as would a red-headed woodpecker. No cliff swallow
personality is manifest. It is obvious why no bird dives at the
boy-intruders throwing stones: there were no boys throwing

stones during the years the cliff swallow personality evolved into subordination to the welfare of the colony. There is no individual in the colony to make the decision. To dive on the stoners would be to dedifferentiate, an event that does not occur with great regularity in things biological, except in those cases where options and choices are required for survival.

Their nests are built of local soil, which is sand, and which holds together in the best of times. The spring rains create very temporary puddles along some of the roads, and there the cliff swallows gather for their nest mud. At the puddle they are more butterflies than birds; their wings are in constant motion, as if they'd had to force themselves down out of the air to the puddle, they appear to weigh nothing, and in the hand they do in fact weigh nothing. The nest itself is described in bird books as "flask shaped." Upon closer examination it is seen to be built of mud balls the size of a swallow's mouth, laid down in rows and layers, roughly concentric and contoured, so that there is a nest chamber and an opening. If this nest is typical of a cliff swallow, then there are genetic instructions within every cliff swallow that tell the bird how to build the nest. The fact of all-consuming colonialism, the fact of an ability to construct a wondrous sand-mud nest in the shape of a flask, are facts of inheritance. They are behavioral traits that the birds have inherited. The advantages of colonialism have resulted in the incorporation of the disadvantages of colonialism into the concept we know as *cliff swallow*.

Off in a museum somewhere there is a "series" of cliff swallows. The "series" is the fifty or hundred or even thousand museum specimens that form the basis for our literary statements and formal definition of the phrase *cliff swallow*. It is from these specimens that measurements are taken, retaken and confirmed, and compared to other measurements of other specimens to determine whether a "separate race" of cliff swallow may be present in California. The series does not reveal the presence of the behavioral genes that determine colonialism or nest shape. Nor are there bedbugs and fleas and fly maggots in the museum series; modern museums take great care to prevent insect damage, and fill their collections with repellent vapors. Rest in peace, cliff

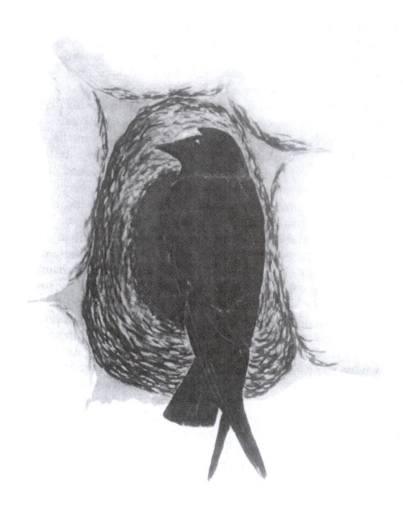

Cliff Swallow
"There is no such thing as a single cliff swallow."

swallows of the series, the tribulations of life have been separated from you in the quiet dignity of the museum research collection.

Back at Lake Ogallala, however, those tribulations continue, and the swallows that control the airspace over the spillway will go home tonight to nests that have not quite made it through the season, dumping eggs and nestlings into the green churning waters of the spillway. Those whose brood is still safe inside the flask will fly directly into the hole and disgorge midge protein into the waiting mouths. It is not known whether an adult cliff swallow is aware of the maggots under the wing of the nestling. *Protocalliphora* is a frightening but beautiful metallic blue fly, which as an adult sucks up liquified dead carp from Arthur Bay, sucks up live cliff swallow nestling as a larva. The maggots are nestled under the nestling wing. The scientific literature goes past this in a hurry, and one has to look quickly to see between the lines. In discussing the life history of *Protocalliphora*, Cole * reviews the experimental completion of the life cycle, noting that a number of *Protocalliphora* species' larvae were reared from nestlings and fledglings of a variety of types of birds. Cole does not tell us whether the nestlings were all alive or dead; but the maggots suck blood, and there is little evidence from biological studies of the last century that dead animals have much circulating blood. The scientific literature is a little weak on one point, however, and that is the matter of whether a single *Protocalliphora* species can be raised on more than one kind of nestling. If one assumes, as one must assume until our hard information is better, that the *Protocalliphora* maggots of the Lake Ogallala colony are restricted to cliff swallows, then there must be a behavioral gene in the fly that brings it to the colony to lay eggs, just as there is a behavioral gene in the swallow that brings it year after year to the spillway to lay eggs. The colonial and faithful swallows, by virtue of their virtues, have acquired a parasite— one who sits beside, that is, at a dinner table—in the form of the fly larvae. The larvae share cliff swallow blood with the cliff swallows.

* Frank R. Cole, *The Flies of Western North America* (Berkeley: University of California Press, 1969).

It is possible to hang over the ledge of the Lake McConaughy spillway, which empties into Lake Ogallala—that is, if someone holds your feet and you have no fear of falling, like the dead swallows, into the maelstrom. The swallows fly like bats into the face of one who peers over the edge into their private world directly under the overhang. One can pick a nest that is occupied but not yet filled with eggs or young and one can break it off and bring it up to the light for close examination. In the light the nest is crawling with bedbugs. *Cimex* is the name of the bedbug, and while the cliff swallow bedbug may not be the same species as the human bedbug, it looks exactly alike to even most biologists. Moreover, the nest itself is covered with bedbug fecal spots. The bedbug population of the colony must be astronomical, and it includes bedbugs of all sizes and ages. There may be a hundred on a piece of nest the size of your hand. The colony extends maybe two hundred feet, maybe three hundred feet, along the edges of the spillway, and occupies hundreds of square feet of the wall of the spillway itself. There must be a lot of bedbugs in the cliff swallow colony. There is no reason to believe their habits are much different from those of human bedbugs: they hide in cracks and come out periodically to suck a blood meal. The walls of the spillway are also covered with bedbugs and bedbug droppings. Contrary to the habits of *Protocalliphora* larvae, bedbugs may also suck the blood of adult swallows, especially at night as they sit on their eggs. The *Protocalliphora* larvae must share cliff swallow blood with not only the cliff swallows but also the bedbugs.

The material that lines the nest can be examined more critically back at the laboratory, where it has been tied up in a plastic bag. In contrast to the barn swallow, the cliff swallow uses mainly grass rather than the feathers of other birds, and the lining is a simple flat pallet. The cliff swallow is a fine flier, make no mistake of that, but the mind stops short when it considers the size of the great horned owl population that would have to be maintained in order to lose enough feathers to line the nests of an average cliff swallow colony. The cliff swallow is a very successful bird, numerically. In the laboratory, under the glare of a desk lamp and the lens of a dissecting microscope, the fauna of the cliff swallow nest pallet is analyzed. Aside from the maggots and bedbugs, there are fleas, both larvae and adults, of the genus *Ceratophyllus*. It is the same genus that sometimes occurs in damaging numbers in or on chicken flocks, and the genus *Ceratophyllus* is generally considered typical of birds and bird nests. Let us see, now, that makes fly larvae, bedbugs, and fleas that all share cliff swallow blood with the cliff swallow. It is not likely to be the bird's personality that these fellow travelers find so attractive, for the cliff swallow has little or no personality. One likes to think it is their blood and their behavior.

The indivdual cliff swallow is the philosophical equivalent of a single cell of the multicellular colony-organism. Their nests abut one another. Their nests are cozy, warm inside, and afford thousands of cracks and crevices for one to hide in. They return to the same colony site year after year. They occur in very large numbers, and hence are successful in the Darwinian sense. They also have acquired the long list of parasites which see the cliff swallow, because of its nesting habits, as a prime target. A behavioral gene—that is, the one dictating nesting habits—has set the bird up for the list of creatures that share its very life's blood.

To the human, viewing the colony and the swarm from a distance, the organism is pleasantly attractive, intriguing. We find some kind of relief that a colony exists on the man-made concrete structures: an indication that the structures may be doing some good other than performing the function for which they were originally intended. In this day and age the human who builds a structure that is both utilitarian and provides space for wildlife is an environmentally successful person. The closer one gets to the colony, however, the less comfortable one is, and to climb down on the divider between the spillway flumes with the colony swirling about one's head is to intrude deep within an organism. You have intruded, quite innocently, into the territory of an organism several hundred feet long, a dinosaur-amoeba. You have no sense of individual birds, you are in the presence of the colony and the colony-amoeba surrounds you on three sides. There is a substantial urge to leave. Hanging over the edge to collect a nest or to net a bird is even more of a shock.

Current biological ideas picture the membrane of a cell as a floating mosaic, a film of fat in which float globs of protein. The surreal magnification of this idea as it occurs in freshman textbooks makes us wonder what it would be like to run into the membrane of a cell many times as large as we. One can imagine the membrane as flowing, with colors not expected, with the globs of protein appearing as mats of hair, sticky, moving and surging, spewing wastes, sucking in the environment of which you suddenly discover you are a part. Hanging over the edge to confront the cliff swallow colony face-to-face at the distance of several inches provides a similar surreal experience. To touch the nest is to touch one part of a naked flowing animal much larger than you, and the animal responds in waves of swallows that fall from the nests and join the complaining swarm. One gets the feeling of having depolarized a three-hundred-foot-long nerve cell with one's finger. It is not a pleasant experience. It is a trip into a world one did not realize existed on earth. It is the strongest experience I have ever had in intruding into the territory of another creature.

The literature tells us that to intrude upon a grizzly bear is sure or near-sure death, and we have come to expect similar kinds of experiences upon intruding into the territory of lions, tigers, and so on. These animals are not like ourselves; their social habits do not produce cities and colonies that because of their internal interactions become the equivalent of single organisms. The cliff swallow does, however, and there is something vaguely familiar about the colony viewed face-to-face. One feels a sense of having been here before but perhaps on another planet, in a dream, in a previous life from which one has been incarnated. Very deep down, particularly in an election year when our faults are magnified by the opposition to the incumbents, every person feels our society has become so complicated that it is being ripped off by a large series of parasites. One senses, also very deep down, that the coterie of friends is there in any society, in any part of the universe. A foot over the edge of the Lake McConaughy spillway is confirmation of that feeling in another part of the universe.

6

Malaria in the Canyons

MALARIA is a disease caused by one-celled animals that live inside other cells of a body, especially blood cells. Malaria is possibly the second or third most prevalent human disease in the world today, but its health is not due to any real supporting efforts on the part of man. To the contrary, humanity has tried mightily since World War II to eradicate the scourge, only to find that the disease-causing one-celled animals as well as the mosquitoes that transmit them tend to evolve faster than human knowledge of drugs and insecticides. Drug-resistant malaria and DDT-resistant mosquitoes are the results of worldwide evolutionary experiments in which chemical compounds were introduced into the environments of mosquitoes and malarial parasites; those creatures responded as populations by becoming genetically resistant. An official concerned with eradication of human disease would say, of the twenty-five-year experiment: "We tried it and it didn't work." The biologist, continually discomfited by lack of hard proof for Darwin's theory, which now pervades all our biological thought, would say of the twenty-five-year experiment: "We tried it and it worked!"

Malaria is a historical phenomenon, having sat in on military staff meetings since well before the advent of the Roman Empire, and along with that other staff officer, Typhus, has made many significant military decisions. Some of those decisions every student now reads about in history books. Malaria is a philosophical

[67]

phenomenon, exhibiting opportunism, showing us a paradox in which high gamble is coupled with high success, teaching us the values of learning to live together with several other kinds of living creatures, teaching us the costs of interdependence. Malaria is an ecologically widespread phenomenon, occurring not only in humans, but also in monkeys, rodents, birds, several kinds of flies, and lizards; and in those species that don't harbor "true" malarial parasites there are very similar one-celled animals called malaria-like organisms.

There are world experts today, making every human effort to relieve us of this burdensome disease, who got their start with bird blood. A bird is held in the left hand, head poking out between the index and middle fingers, ring finger and thumb holding the left leg, while the right hand plucks the feathers off the inside of the left leg. At the tarso-metatarsal joint there is a blood vessel flowing between the leg bone and the major tendon. The vessel is pierced in a flash, blood wells up instantly, the bird jerks in your hand, the drop of blood is smeared carefully on a glass slide, and the bird is released to hustle back into the junipers. You were the bird the last time you went to the blood bank to donate; the only difference is you were not allowed to run off as rapidly as your legs would carry you into the neighbors' trees. The bird is gone, basically unhurt, the bleeding will stop very quickly if the bird is released quickly, and you've caught too many with legs and toes missing to ever believe your drop sample is going to have any effect on that animal. Later you stain the blood smear, just as your blood was stained the last time you went in to donate blood or for a physical checkup. However, in the blood cells of this bird there are amoebalike animals. They stain a blue and red that is instantly recognizable. Old friends. The infection is alive and well again among those canyon birds.

The mist net is the basic tool, and there is a ritual that accompanies the setting of the net for the first time each year. Tribesmen have been setting nets for untold centuries; tribesmen have ritualized their hunting techniques in thanks to various deities and in dances, and we are not very different standing in the canyon with nylon mist net. The ritual is about to begin. The air is charged. Two stalwart lads assemble the poles; the women deftly string the loops, tie the net loosely in the middle to prevent tangles; and others gaggle and joke and stomp down the poison ivy and break off dead hackberry branches and smoke cigarettes as the stalwart lads begin their clamber down the canyon sides with their poles and the strung net. The middle ties are pulled, and the net falls across the canyon, blocking a natural flyway for canyon birds. There is an "ooh!" and an "ah!" as the net falls into place. The stalwart lads survey their work with a smile. The women are apprehensive, yet hunger more for the actual bird than do the lads. And the first drive begins.

The tribe slips softly into the junipers and up through the Brule rocks and yuccas toward termite country on top of the bluffs. It is windy on top of the bluffs, and they assemble in their sweat shirts and blue jeans, for it is cool this early in the summer. There are some people on the bluffs who will before long receive a doctoral degree from a major university. There are some people on the bluffs who have never made any grade other than *A* in school. There are some people on the bluffs who will leave soon for advanced scholarship research in Europe, and there are plenty of people up on the bluffs who will be family physicians and surgeons, actually cutting a live human body, in some few years. Yet today they are a tribe, waiting for the signal, and on that signal they leap yelling and waving their arms into the canyon, down the outcrops, through the tangled junipers, pushing the canyon birds hopefully into the net two hundred yards below. Thus the first drive begins.

The first drive often nets nothing, but it is the beginning of the netting season, several weeks of daily tribute to the mist net. Today it has netted nothing, although tomorrow it will net a full adult magpie, but the drive's failure has no effect on the spirits of the tribe. They continue along the road beneath the bluffs, toward the rock dam and white water where the blackfly larvae live. The net is up and it will work sometime. One year the first drive netted something: a whole family of brown thrashers. Mama, papa, and four baby thrashers—fledglings, really. They all hit the net and stopped, hanging upside down, fluttering.

[69]

The tribe worked quickly, jabbering, to free the family. The entire family had malaria. No surprise as far as the adults were
concerned, but for the fledglings it was startling. Typically it
takes ten days for the infection to become patent, to actually
appear in the bird's blood, and these fledglings were not much
more than ten days out of the egg. Six infected birds in one net
is also a hell of a lot of infected birds, especially on the first
drive, and especially when you might expect six out of a hundred
at any randomly selected location. Those fledglings might have
been bitten as soon as they hatched, bitten by the flies transmitting
the parasite. If so, there was one hell of a lot of natural transmission of bird malaria in those canyons. You stood with the birds in
your hand, ready to release the family, and stared up into the
junipers. The wind whistled up through the canyon, around the
rocks and through the trees and tangles, and you had this vision
of a mosquito sitting on a brown thrasher egg while the shell is
pipped. It doesn't take much of a hole for a mosquito, just a
fraction of a square inch, and the thrasher's beak is scarcely showing through the hole when the mosquito slips inside the egg for
its first blood meal in a week. In that week the malarial parasites
it picked up from the last bird it bit have matured, wormed their
way into the salivary glands. Now they flow into the thrasher
chick, blind, still in the egg and fighting hard to get out. Yes the
infection was alive and well again that year in the birds of the
canyons.

A bird's red blood cell is an oval with a raised center, the
nucleus. On a slide the cell stains a light orange and the nucleus
stains a dark purple. A bird malaria project requires the careful
study of hundreds of stained slides, and even years after such a
study, a bird blood cell remains a major nostalgia trip. Through
the microscope an old man can go right back to that time with
the thrasher or meadowlark in hand, and disappear among the
cells as he stoops from one to one looking for that blue amoeba
nestled beside the nucleus. The cells are superenlarged, and as
the old man falls into the microscope tube they swirl about him
and bump out into the room, bouncing bubbles from one of those
giant kid's loops. The surface of those cells is multicolored,

kaleidoscopic, a throbbing light show, with large globules of twisted proteins floating, sticky, some submerged hippolike in the greasy membrane, some riding high on the color waves, corks, pilgrim ships. The old man feels a breeze, then a total environment current as the cell sucks in its surroundings. Now the breeze and the currents around his feet reverse, and the fetid exhalation of the cell spews sharply pointed molecules at him, some acid and stinging, some sticky, and some winding about his arms and tangling his hair. He places his hands on the hot membrane and rubs the colors away, holding the slippery cell against a corner. Peering inside he sees the blue amoeba, moving now in globular waves, always keeping in the end between the nucleus and the tip now firmly wedged into the corner. There, inside that amoeba, are the red balls! Suddenly there are more, and before the old man can get his act together the amoeba has split into several, and the bottom has dropped out of the red blood cell. Hemoglobin gushes out and over his ankles, and the little amoebas nip his legs as they ride the crest between them. They are gone now, mongrel pups yapping at this cell and that until in a daze down the hall in the fading light the old man sees them inside other cells. There are two in one now, and he wonders if they'll get along. His eyes lift from the microscope, and the fluorescent lights return the dingy room to normal. He takes a plastic pencil from beneath a sweater, pushes several times with his thumb on the silver top until the microthin lead emerges, then writes in careful printing, "Infected."

Day after day and week after week they parade through the net, and an incredibly high percentage have the disease. Some have true malaria; some are infected with malarialike parasites, still as blue amoebas within the red blood cells. Some hit the net gently and struggle gently, and those are freed quickly, for the net is checked several times a day. Some hit this net and struggle violently, and sometimes those don't live. Then there are some, such as the full adult magpie, that hit the net with a vengeance during the storm that rocks the lake and canyons and throws the net into the highest hackberry limbs. There are red-headed woodpeckers, flickers, chickadees, red-winged blackbirds, yellow-headed blackbirds, magpies, blue jays, grosbeaks, orioles, swallows, and kingbirds, and they all have something in their blood. Back in a dusty library a person going through the old Dewey literature, generated during the heyday of bird malaria, forms an image of his canyon in Keith County. There are more malaria and malarialike bird infections in that one Keith County canyon than in the whole of most other nations. The person returns to the canyon and again stands staring on the gravel road while the wind moves up through the junipers.

It is just a small canyon between a couple of bluffs. The sides are steep in places, and there is one spot where an owl has roosted, a large overhang where a fire could be built and an evening spent listening to that owl hunt. It is just a small canyon filled with juniper trees of all sizes, junipers and hackberries, a little poison ivy, a lot of yucca on the exposed sides. It is dark in there, and tangled in places, and there must be tangles where a thrasher nests, for above all else the brown thrasher in that canyon symbolizes malaria. There is something special about that canyon. When every brown thrasher in a canyon has an infection, sometimes with two different kinds of malarialike organisms, then there is distinctly something special about that canyon. A dozen brown thrashers in the hand in the last week or two, and all were infected. The infection is alive and well, certainly, but the tightness of such an ecological relationship between bird, protozoan, and fly goes well beyond that allowable by current biological science. There is much chance in this relationship: the fly must hatch and live in large numbers, the fly must bite the infected bird, the infected bird must have the blue amoebas in its blood in adequate numbers at the time the fly bites, the fly must live long enough to get hungry and bite another bird, and that other bird must be susceptible, if not a brown thrasher. Even the chances of a blood-sucking fly living long enough to get hungry are infintesimal. Yes, a canyon where every thrasher is infected is quite a canyon indeed.

We must get technical. There are three major kinds of malarialike diseases of birds, each caused by a different kind of protozoan. In nature the diseases are generally not serious, at least we don't

Brown Thrasher

"It is dark in there, and tangled in places, and there must be tangles where a thrasher nests, for above all else the brown thrasher in that canyon symbolizes malaria."

think they are; so they are not called diseases but rather "infections." All of these three types of protozoans spend some time circulating in the bird's blood, within the red blood cells, although the life-cycle stage of the form in the red blood cells is somewhat different in each case. There are a variety of flies that transmit these infections, and all suck blood—bird blood. To date there is no documented case of a human getting infected with a protozoan that properly belongs in a bird; but then who knows?— monkeys get infected with human malarial parasites.

Mosquitoes transmit malaria, and mosquitoes are flies. The other bird blood infections are transmitted by other types of flies, and there are several. Blackflies, or buffalo gnats; no-see-ums; and mosquitoes—all bite birds, and to the initiate half the business of bird malaria is flies. To the veteran, well over half the business of bird malaria is flies, for a bird can be captured, shot if necessary, raised from an egg, fairly easily located, its blood sampled hourly for days or daily for weeks; and the warm-blooded and intelligent bird does for you what you cannot do yourself: it sustains the infection if it has it. The fly, on the other hand, is difficult. Proof that a fly is actually a vector in nature is a very difficult task indeed; it is exceeded in difficulty only by the task of proving that a fly is the most important vector in nature. Yes, well over half the business of malaria is flies. If flies were not so beautiful, so endearing, the fly business would be onerous. As it is, the fly business usually takes over quickly from the bird business in one's mind. Mosquitoes are easy. Black flies are not bad. No-see-ums convert this serious scientific business into a joke. The Keystone cops chase no-see-ums. They'll try it this next year, because they got a sample of cloth from Canada. One needs a special sample of cloth from Canada to chase no-see-ums, not too unlike the special cloth from which were made the emperor's new clothes. Bird malaria in the field is contraption city well beyond one's imagination. There are people in the world who make a living at this and are happy with it.

There was a time in the midde of Kansas when the cattails were parted down in the ditch by the gravel road, and the still water lay an inch or two deep over the mud, wet so many times the last weeks and months. The sunlight filters in at these times, and strikes the water clear to the bottom with its vegetation litter. The water is very clear, and the brown and wet cottonwood leaves of last year are seen to be crawling with microanimals. Sometimes if the day is still one can even see protozoa, especially the large blue ones. Incessant activity in that water over the leaves is the mosquito larva population. All sizes they come, all wrigglers, with hefty shoulders, raking mouths, spines and brushes on their air tubes. They are captivating in a white porcelain pan; they move about the surface, jerk to the bottom. Sometimes there is a pupa. It is all a wondrous sight and you forget they are mosquitoes. Back in the laboratory they emerge as fresh adults with white banded legs. Some have fallen into the water in that last attempt to clear the interface between pupa and adult, and you feel very sorry for them. During the night some made it and some didn't, and the didn't ones lie in a pool of feathery scales, rapidly growing fungus. They drink the sugar water you give them and they fly around the screen cage. Pets. Your mosquito pets. Two or three years of mosquito pets and you will never see another mosquito without a soft spot in your heart. You will always stop and look at a mosquito. Once around the light trap you saw some *Psorophora*. They were almost an inch long, and they were the most beautiful mosquitoes you'd ever seen until you saw the *Deinocerites*. Out in some tree hole full of water the *Deinocerites* female had laid her eggs, and out into your light trap came the black youngster with glowing silver lines. That's the mosquito business. It is a shame that as species they are so skillful at the malaria and yellow fever games . . . humanity's image of the mosquito is all distorted. They're just trying to make a living. There are few mosquitoes in Keith County. The larvae are not there in every roadside ditch, and one must look long and hard for the old friends. In Kansas they are everywhere, even thousands of acres of *Aedes* larvae after a June storm floods a pasture. In Keith County they are tucked away in backwaters until you put up a baited trap. A few yards up on the bluff you put a large contraption with a screen cone in one end and a cloth sleeve over the other. A bird goes inside, in its own little hardware cloth cage. It is morning now, and cool, and you stomp the few yards up to your trap in the junipers. It is filled with engorged mosquitoes. The bird is released. Yes, the infection is alive and well in the canyons.

The blackflies are out on the rocks. In Keith County they're on the rocks that hold the water where the state record rainbow trout was caught, they're out on the rocks where the water from Big Mac boils out of the toe drains, and they're out on the concrete where the North Platte is sent down to Sutherland. The adults are rarely seen, but larvae are clinging to rocks by the millions. They are panned up along with the snail eggs, but the larvae belong there. They are marvelously equipped for belonging there, with bulbous rear ends and webs that hold them to the rocks. They romp in the white water and love it. Some day they'll bite a bird.

In Keith County it is safe to stand by the river and watch the blackfly rocks. There are places in the world where one could go blind looking at blackfly rocks, for the blackfly transmits another kind of infection in such places, an infection with a worm, *Onchocerca volvulus*. You have stood on the banks of Keith County fly streams for two years, and have yet to see an adult, although the rocks and concrete ditch walls are black with the blackfly larvae. Out of all the quadrillions of insects you have seen in two years, you finally get bitten, once, standing where the toe drain dumps into Lake Ogallala. You capture the thing and put it in a small vial, a now supervaluable small vial. You try to identify the adult. The adult makes you weary. You knew it all along but were just too stubborn or lazy to admit it: bird malaria in the canyons, especially if complicated by the blackfly transmission of some malarialike organism, is now a twenty-year job. You have already spent two of the twenty, only eighteen to go to understand the ecological, behavioral, social, and climactic factors that regulate the intensity of malarial infections in the juniper canyons. The canyon is a microcosm of vector-borne disease. The brown thrasher is a model; its blood tells you of associations between species that you know will take twenty years to explain. Another year of the twenty will probably be spent identifying the one adult blackfly. But there's more.

No-see-ums are small flies of the family Ceratopogonidae. After a little practice, the word *Ceratopogonidae* becomes several times as easy to say as *no-see-ums*, so everyone simply calls them *ceratopogonids*. The word is even fun to say sometimes. Inevitably some person bridles at the use of a scientific name. Use, however, makes them fun, especially if one starts with names like *ceratopogonids*. The word has a nice ring to it. Now try *Culicoides*, for *Culicoides* (*Cūl-ĭ-coid'-ēs*) is the generic name for the most important group of ceratopogonids. Repeat this last phrase several times. You are now a biologist working on the most difficult and surreal part of bird malaria.

A former student comes to town, a former student made good up in Canada, where he is a resident expert on bird malarias. I am exceedingly proud of the former student, for he has distinguished himself in all the way I knew he would. He has also come back to town to tell the old man a thing or two. Former-student-made good-in-Canada cannot come to town without looking at some brown thrasher blood, and he sits at the microscope shaking his head. The thrasher is infected with at least two kinds of organisms, and he shows you specific details you have overlooked on that slide for two years. The infection is more alive and well in the canyons than you realized. In fact, the infection may be so alive and well that it is well beyond you and your sophomoric abilities out in the junipers! Talk turns to *Culicoides*. The former student brings out some two-by-two slides, and you all sit in the darkened room learning from the screen the techniques for studying bird malarias in the broadest sense. You shake your head at this. Things were out of hand anyway, in the canyons, with the mist nets, mosquito live traps, hardware cloth cages, the vial with one adult blackfly, thrice-daily homage to the net, blood, slides, microscopes, assistants, blue amoebas—all out in the canyons. Now this guy tells us we need a chicken on a platform and an insect trap we can crawl inside. Contraption city, out there in the junipers. One does not ignore the advice of an expert. Next year the Keystone cops *will* take a chicken to the junipers and hoist it into a tree. It remains to be seen whether they will be quick enough to get the chicken down from the tree the next morning and get the special-cloth insect trap over the chicken in time to catch an animal with a name like *no-see-um*. The name itself should put one on guard. Instead of the mist net ritual, maybe this bird malaria business should begin with a snipe hunt. We could then use the snipe to hoist into the tree to catch an animal with a warning name like *no-see-um*!

There are times now when I think about the twenty-year malaria problem, and these are the few times when the malaria business does not reach for me a state of intellectual collapse. It is obvious that one goes to contraption city out of love and joy, for as in the *Fundulus* work, there is no money or pride to be had in bird malaria. There may be some pride, but the Keystone cop learns early not to have too much pride, for pride alone will not sustain twenty years of research on the social activities and close associates of the brown thrasher. There are no brown thrashers wild in India, for the Mimidae is a New World family of birds; so Ronald Ross could not have done his pioneering malaria research on brown thrashers. Even in the late 1800s there was conflict over the rights to Ross's discovery that malaria was transmitted by mosquitoes. Sir Ronald Ross is pictured in India in the 1890s as a towering mustached man standing on some steps with his native assistants and cages. His original writings speak of sparrows and grey mosquitoes, and by the early 1900s he had won a Nobel prize for his discovery that malaria was transmitted by mosquitoes. This man was a world figure in medicine at the time, so he wrote his memoirs and humbly called them something like *The Great Malaria Problem and How I Solved It.*

It has only been eighty-five years now since Sir Ronald Ross "solved" the great malaria problem, and although we know the infection is transmitted by flies, we still have not come overly close to understanding the factors that regulate the level of such a disease in host populations. Before the no-see-um incident I had given myself twenty years in the juniper canyon just east of the boat house. I now find myself standing in a room stacked to the ceiling with mist nets, mist net poles and ropes, large cans, screen cones, white pans, cloth sleeves, birds of all kinds, microscopes, cages, hoists and platforms for getting the chickens up into the tree, chickens, trees, the tribe waiting for a signal, mosquitoes swarming about my head. In my hand is a sample of cloth from Canada; it is the only kind of cloth that can be used to make a no-see-um trap, and the no-see-um trap has to be large enough to be quickly put over the chicken, out of the tree on its platform, and has to be made so that a cop can go inside and start looking for a thing with a name like *no-see-um.* We are ready and poised for the malaria season still several months off. I hum a tune, standing in the room poised and ready, and it's a tune from Waylon Jennings.

> Tell me once again so's I'll understand,
> are you sure Ross done it this-a-way?
> Did ole Ross really do it this-a-way?

7

Grasshoppers and
the Ackley Valley Ranch

IT WOULD HELP, if one is going to write about grasshoppers, to actually know something about them. Every boy knows a grasshopper, just as every boy knows a grasshopper spits tobacco on your hand, but few apart from professional entomologists, grasshopper freaks, and county agents really know what a grasshopper is, and I am not an entomologist, grasshopper freak, or county agent. I will become a grasshopper freak, however, within the next few years, and that knowledge carries an excitement, an anticipation. Perhaps even more exciting is the knowledge that some day within the next few years a student will walk into my office and we will talk about grasshoppers for a while and then that student will walk out, go to the Ackley Valley Ranch, and become a grasshopper freak with identification skills far in excess of mine, and a very few years thereafter make his or her own scientific reputation on his or her own analysis of the animals that live inside grasshoppers.

There is smugness in knowledge like that. There is smugness in knowing that a valuable jewel lies in the grass of the Ackley Valley Ranch, and all someone has to do is go pluck it up with an insect net. There is a smugness in knowing that thousands of others could walk those fields day after day and never see that jewel; its security is in its simplicity and obviousness. It is a jewel of knowledge and discovery that will make the difference in the life of some human, because for that human it will become a life focus, it will give that human a raison d'etre, it will give that human something no society or tyrant can take away. At present it's nothing more than what is often called "a problem" in this business, it has no immediate economic significance, its solution will never be used to cure cancer or heart disease, and not even the Ackley Valley Ranch will realize any direct "benefits" from the solution, as if Waldo Haythorn needed any of the things we usually think of as benefits. It will however turn that student into one dyn-o-mite ecologist, a person with a view of the world that others will value and want to follow.

The twenty-five-year prospects of the Ackley Valley Ranch grasshopper problem are so exciting it is difficult to write about them. The urge is almost too strong, the urge to simply get into the car and drive to the Ackley Valley Ranch and begin work on the problem today. Fortunately for the city obligations it has become so cold in Keith County that the grasshoppers are gone until spring. Fortunately also for the city obligations the ultimate thrill in this business is not in the doing, but in the guiding of others' doing. I could work on that problem easily myself, out of love and joy, with a minimum of expense, and publish several scholarly works. At the end, there would be some pages in some learned journal with my name on them. At the end I would have accomplished for the world only a fraction of what was possible had I spent the equivalent time convincing some student smarter than I to "take the problem" as his or her own.

We were in a hotel lobby in Kansas City at some large scientific meetings when I was approached by an older but decidedly dapper and vibrant gentleman. There is no way to write the tone of his voice, but it could be partially described as raspy-devilish-blustery-Irish.

"Anybody ever tell you the secret to success?" he blustered.

"No, Dr. Coatney, tell me the secret to success."

"*Surround* yourself with people smarter than you are then *get* out of their way and *let* 'em work!" He followed this with a devilish blustery Irish laugh, then stabbed his index finger into my tie. "Then *you* go out and talk about what they did!" It is almost unfair to the rest of humanity for one person to have the

knowledge that eventually he will be able to talk about what some
smarter student did with the grasshoppers out at the Ackley Valley Ranch!

Ackley Valley is owned by the Haythorn Land and Cattle
Company, and Waldo Haythorn is the patriarch of that organiza-
tion. Headquarters are near the town of Arthur, in Arthur County,
several miles north of Arthur Bay. There is an art gallery in
Arthur, and a city park bordered with very large cottonwoods.
A person normally lowers his voice when entering the town of
Arthur. There is no local ordinance against loud talking or cursing, but ten feet around that last curve of the highway into town
there is a feeling that a person should be quiet, that one is in a
very special place that belongs to someone else. It's that kind of
a town, especially on a very hot midsummer Sunday afternoon.
Maybe the fact alone that Waldo lives there causes one to lower
one's voice in respect for the place. One also has total respect for
the man who owns the Ackley Valley Ranch, even if that man is
only a legend to you, for the man who owns the Ackley Valley
Ranch also owns the long-billed curlews and the great blue herons
and the box turtles and the grasshoppers of Ackley Valley, and
that man has given you access to the ranch. That man is also said
to cut hay with a mower drawn by Belgian horses.

Our gate to Ackley Valley is north of the place they call Sportsman's Complex, north along the road to the town of Arthur, north
along a road that gives one a tingle the minute one passes the
complex, north into a country that you know is wild, different,
uninhabited, north into curlew country, north into Haythorn country, north along a road that you can drive and drive and drive
and drive without ever seeing a sign of city civilization. The pick-
ups you meet coming south always wave with the index finger,
raised slightly from the steering wheel. If you don't know to look
for an index finger raised an inch above the casually held steering
wheel of an oncoming pickup truck you will miss a greeting. Miss
a greeting under these circumstances and you're immediately
branded as a stranger. Pretty soon you find yourself giving the
greeting to every car you meet. Sometimes after a few beers back
in the city you find yourself giving the greeting to all the cars
in the city traffic.

Great Blue Heron—Ackley Valley Ranch

*". . . the man who owns the Ackley Valley Ranch also owns
the long-billed curlews and the great blue herons and the
box turtles and the grasshoppers of Ackley Valley. . . ."*

The Ackley Valley gate is a few hundred yards past the cattails, and much of the anticipation of Ackley Valley is found in the cattails. There is standing water near the road and acres of cattails, and in the distance there are large cottonwoods, and different shades of grass indicating some very wet areas. The Ackley Valley gate is done right, with a closer made from what looks like branding iron parts. The first time out, Carla, the country girl, whipped the gate open instantly and closed it behind us instantly. The complexity of a Sandhills gate-closer was not apparent until the doctoral student from Connecticut tried it. There is a nightmare to be had while sleeping in the Sandhills. In this nightmare you are riding with Waldo Haythorn in his pickup truck and come to a Sandhills barbed wire gate, which out of courtesy you jump out to open only to find you can't work the closer. You try to cope.

Ackley Valley is curlew country. People come in contact with birds like curlews in storybooks, maybe in zoos, never on city streets or in city parks, and decidedly never in their Sandhills moods unless those people step inside the Ackley Valley gate. There are other places, such as the Crescent Lake refuge, where one can expect to meet a curlew *some* time during a day, but there are few places like the Ackley Valley Ranch where one knows one can meet a curlew *any* time during the day, and *all* the times some days.

A long-billed curlew is a mystery bird, the kind of bird that might appear in a Disney real-life adventure film, the kind of bird that makes films like these successful, because it is so far outside the daily experience of every person. Few people ever have their hands on a wild bird, and even then it's likely to be a duck, quail, or pheasant killed while hunting, a bird to be converted into refrigerator meat at the earliest possible moment. A friend carves decoys, and on a trip to his home I discovered a long-billed curlew, standing by the fireplace, carved perfectly, in stately posture. The carving reeked of shorebird, the bill screamed sandpiper, the pattern disarmed me the way a yellowlegs or long-billed dowitcher

[83]

can disarm me every time. There by the fireplace was the Ackley Valley Ranch.

There was a time when I thought those contacts with dowitchers and yellowlegs on the Gulf Coast and on the plains of Oklahoma and Kansas gave me a part of nature that only so few of the other four billion people here had. I still feel that way, partly, although the memories of those birds are fast fading; somehow the mudflats necessary to supply me with dowitchers and yellowlegs are not here in time or place. I have a friend, however, who does research on curlews. We communicate instantly, better than I do with the decoy carver, and we communicate without speaking on many matters that extend far beyond the biological business at hand. We can read the academic politics of the day in each other's tone of voice. It's because we've both seen the long-billed curlew, although he obviously has seen it more closely than I. One step into the Ackley Valley Ranch and you meet the curlew.

There is no search, no effort required, only your physical presence. The curlew brings it to you full force. The curlew throws off all the fear and seclusion you think should be a part of the personality of so large, so beautiful, so dramatic, and so oh-so-vulnerable a bird in a world shaped by the Bureau of Reclamation. The curlew screams at close range, the curlew sets its wings at every angle you've ever seen in the most impressive pictures, the curlew dives, it circles, it comes back for more, and all you have to do is wave your arms and it does everything all over again twice as powerfully. It hardly helps to leave.

The curlew follows you down the highway screaming, the curlew flies in front of your van looking over this shoulder, then that, staying just far enough ahead, as you pick up speed, to scream back at you. Down the highway a couple of miles you discover you've hardly been breathing. There is something special about a place where there are *always* curlews. Even if you never meet the grasshoppers of Ackley Valley, the curlews have given you far more than you bargained for.

The total lack of fear in an animal that should above all be fearful of what humanity has to offer is an impressive demonstration. It makes one wonder long and hard about what is to be gained by fear, especially what is to be gained by fear of forces that should be great and much beyond our control. I have some very serious doubts that the long-billed curlew would allow an earthen dam to be constructed over fissures upstream from where it nests. I have very serious doubts that the long-billed curlew would allow a nuclear waste disposal facility to be built on the Ackley Valley Ranch, regardless of how much the curlews needed the jobs. I have this almost uncontrollable urge to find every human on this earth who has ever feared to think about and to conclude, speak up and out, and act on environmental issues and lead that human by the hand to the Ackley Valley Ranch in June. I also have this feeling that the curlews approve of a man who is said to still cut his hay with a mower drawn by Belgian horses.

The grasshoppers are everywhere, and in early June they are mostly small. They rattle in the hard grass as you walk, unseeing, toward the cattails and beneath the curlews. They rise in waves and blow like bits of leaves along the tops of the grass blades. Even as they bounce off your jeans you are more interested in what is underneath the cow pie than what is underneath the exoskeleton of a grasshopper. It is significant that we have tramped Ackley Valley for two years before turning one day in frustration to a netful of grasshoppers. A preconceived notion of what there is to be seen has led a too-smug and too-conditioned invertebrate zoologist to the cattails. In retrospect, the world-renowned scientist had more to say to us than we realized. He showed up on invitation, a human form for a name that had been branded in my brain so many years ago, a hero of the kind you are not always sure really had bodies to go with the names. We tried to get him to the cattails and he refused. He wrote them off as too complicated to deal with. This came as something of a blow: out on the prairie one heads directly for the cattails and wades in, for there are to be found the least bitterns and the yellowheads; and in the Sandhills, the wrens. On the other hand, when your hero refuses to go into the cattails you stop and think, and to stop and think about the cattails of Ackley Valley makes you realize that you've walked directly past the grasshoppers in your drive to the

[85]

Practice Curlews

*"The curlew throws off all the fear and seclusion you think
should be a part of the personality of so large, so beautiful,
so dramatic and so oh-so vulnerable a bird. . . ."*

cattails, and what you found there will never in any way match what you found on that day of frustration when you caught a netful of grasshoppers.

A minute, let alone two years, in a marsh is never wasted, however, for there are large snails of the genus *Helisoma* in the Ackley Valley cattail marsh, and you are not seeing *Helisoma* anywhere else, and not only that but *Helisoma* is a first-rate nostalgia trip. *Helisoma* has a flat spiral, and in the Ackley Valley cattails it is very large, robust, and healthy. There is an urge to simply pop one in your mouth, standing knee-deep in the typical marsh muck. Later you realize you should have given in to the urge; there are no parasitic worms in *Helisoma* livers on the Ackley Valley Ranch. The snails are not only healthy, they are obnoxiously healthy. Out in the fields away from the marsh there is a very wet spot where spring rains still linger and chorus frogs lay their eggs. You pan the wet spot looking for one-celled animals, and instead you find another flat-spiraled snail, this one of the genus *Gyraulus*. Even the adults are very small, very delicate, and their bodies can be seen through their shells when viewed with a dissecting microscope. *Gyraulus* is beautiful and delicate, but beauty and delicacy are no defense when it comes to worms, and through the shell one can see not only the snail's body but also the parasitic worms jerking within that body.

Two flat-spiraled snails in *Stagnicola*-land is a true treat, and they instantly brand the Ackley Valley Ranch as a different place, an unusual place. Their presence confirms what the curlews have told you: step into the fields through the branding-iron-scraps gate-closer and you have stepped into Haythorn country, where some creatures who refuse to show themselves in the public of Arthur Bay have found a refuge. Maybe it's the Belgians. One doesn't even hear about Belgians much around here anymore, much less hear about them being used for haying. Maybe in a few years the numbing repetition of scientific research will have taken the shine off the Ackley Valley Ranch. Maybe there will be several, even many, places where the curlews shame you every day with their fearlessness, maybe there will be found miles and miles of streams and marshes and potholes where *Helisoma* and *Gyraulus* are healthy and parasitized and abundant. Maybe we will hear about several people who still cut hay with a mower drawn by Belgians. Maybe there will be many places discovered where binding electricity flows through every blade of grass and every cattail stalk and every cottonwood tree into the bodies of every animal that ever ate a blade of that grass or ever found a home in the cattails. Maybe there will be several places discovered where that binding electricity, that sense of Waldo Hawthorn's immense respect for the land, produces in the inhabitants the confidence to be visible, up front, up on the table, not afraid, not afraid to be found, not afraid to put their best foot forward whether it be the spindly but wiry foot of the curlew or the slimy foot of *Helisoma* or the oh-so-functional foot of the grasshoppers. There is this feeling, standing in the Ackley Valley cattail marsh, that if Waldo's respect for this land were extended beyond the borders of his ranch and into the cities, there would appear all sorts of things no one realized existed. There is this feeling that if a person mowed his back yard with Belgians, then *Helisoma* and *Gyraulus* would appear in the kitchen sink, and that if every person mowed his back yard with Belgians then the curlews would fly along the city streets screaming at the traffic.

Standing knee-deep in the muck of the prairie cattail marsh you feel the Haythorn binding electricity soaking through your jeans. There is a force in this place, a force that communicates with every living thing, that establishes the psychic environment encompassing snail and curlew and human alike. The force draws you back to Ackley Valley even when you have no reason to go there and even when you know exactly what you will find. There is no place except within the Ackley Valley barbed wire gate where you feel that force so strongly, where you feel so in tune with the frequencies of snails and curlews. It is a force that gives you strength, and although it is a force that puts you on a level with the snails, it is also a force that puts the snails on a very high level indeed. The sense of Ackley Valley is the sense of togetherness, of confidence in associations, of willingness to be vulnerable, to be different, and most of all to act from a basis of that confidence and willingness. What better place to turn to in frustration with one-celled animals? I should have known weeks earlier that the grasshoppers of the Ackley Valley Ranch would provide the best lessons on the ways living things encounter one another and maintain an association. I should have known weeks earlier that not only would they have provided the best lessons, but that they would have taught those lessons so fast that I would be inundated. Typical Ackley Valley biology: turn to an Ackley Valley animal and that animal gives you back more than you bargained for faster than you bargained for it.

There are things that people do periodically, out of habit and in answer to some call. These are things that people don't go for long without doing, for if they go for very long, then there is some call and off they go to do this thing. I don't go for very long without looking for a one-celled animal. Normally it happens every day, often several times a day. By requirement it happens once a week, and their appearance through the microscope assures me that all is well ("Week of the 22nd of November and all is well"), that they are growing in the glass host I have provided. The weeks in the field are not much different. A person knows they are out there, a person drives along a cattail marsh and knows the dead vegetation will produce a standing crop of one-celled animals. A person sees a forty-pound snapping turtle and simply *knows* that one-celled animals are living within the blood of that turtle. A person seines up a hundred little fish out of any plains river and simply *knows* that the one-celled animals will be there on the gills. A person should also know that the field of grasshoppers is, in addition, a field of one-celled animals. They are called gregarines, are transmitted by spores, and therein lies the tale.

A gregarine in a grasshopper intestine is a memory of an encounter, and several gregarines in an intestine are the memories of several encounters. One species comes in direct contact with an unrelated species, and the tracks of that encounter remain in the form of one-celled animals in a grasshopper's intestine. It isn't enough simply to find a one-celled animal: we find plenty of them. The pressure is on to find not only the animal but the context, not only the animal but the obvious significance of that animal's position in nature, and the pressure is being applied by those who have already found *their* animals and contexts. That kind of pressure can create some frustration and ultimately some disgust. The disgust is with oneself for not acting on the information at hand, not committing resources to an action when one *knows* that action will yield the desired results. I have only known for about twenty years that grasshoppers have gregarines, and I am wondering now why it took so long to sweep up that first netful on purpose.

The gregarines are there, in the intestines where they are supposed to be. But there is more: the grasshoppers of Ackley Valley are togetherness itself, for they are a veritable community, with other insects' larvae in their bodies and external parasites of all kinds. The grasshopper's world at Ackley Valley must also be one of intense competition: one sweep of the net yields several kinds and several ages of individuals of each kind. It doesn't take long before one realizes that a simple idea like cutting up a few grasshoppers to find some one-celled animals has suddenly produced a set of questions that can form someone's lifework. I should have known long ago that the gregarines would be in those intestines; a curlew would never have blundered along for weeks without acting on that knowledge. I conclude that an analysis of grasshopper populations and the factors that regulate their gregarine loads is really an analysis of the way two populations of unrelated animals come together on this earth and maintain an association, an association that is absolutely essential for one of them and a burden on the other. I am confident such an analysis is worthy of being someone's life-focus. I have no hesitation to flaunt that confidence, to throw it out to students, knowing full well that one of these days, one of these lost and lonely days, that right student will come along, focus on the grasshoppers of Ackley Valley, become a grasshopper freak, a dyn-o-mite ecologist, and will spend the rest of his life teaching values and approaches to other humans on this planet Earth. After all, the association between human and Earth is one that is absolutely essential for one and a burden on the other. The curlews are screaming overhead now, and I sense they are telling me to get on with the task, to find that person, but the right person, and find him or her quickly; there may not really be that many Ackley Valley Ranches left.

8

The Lady Whitetail

W E HAVE APPROACHED nature and we have asked ourselves what there is to be learned from nature, rather than about nature, out in these hills of termite country. We have chosen our places to find animals without really knowing *why*, other than that they were easily accessible at the time. A field expedient decision simply to walk down to the Keystone Marsh rather than drive fifty miles to Crescent Lake, or simply to drive to Arthur Bay because it's the first fairly noncivilized spot where there's a road along the north shore, these are the choices that led us to those places the first times. Late at night in Bill and Rhonda's restaurant, or around the fire in the lab, or even back in the city in midwinter, those places are always referred to as places, although the talk moves quickly to the creatures found there. Even in Keith County, even on a research morning, a class morning, we always choose a place. Maybe it is simply habit; if a human is to go somewhere, it may simply be habit to mention the somewhere, leaving to the imagination the activities to be conducted. So rarely do we mention the activities to be conducted and leave to the imagination the place.

There is something at work in my mind now, having done this for so many years, having assumed that if one goes to a place, one will find things there to be studied, that says maybe the place itself has as much to say as does the nature found there. The biologist approaches nature in the form of a plant or animal and

[91]

immediately begins asking questions about the innermost soul, the innermost characteristics, the true spectrum as well as the immediate traits, of the living thing. Even one biologist to another, even one human to another, when thrown together, will begin that searching through the layers upon layers of things that spell that person's name, until sometimes on purpose but sometimes by accident an innermost layer will be found.

In our search of other humans we find that some humans have been reduced to mere repositories of the cast-off things of others —dumps, places that we thought were people at first. We find still others who have an inner character that directs their ultimate form regardless of the alterations of immediate environments, and often one must search long and hard for this inner character, with the inner character being revealed only after a long progression of experiences, each peeling a layer of previous experience, each putting a previous experience into a somewhat different light, so that in the end the true nature of the beast is revealed. One would like to discover the true nature of any animal, if one is a zoologist, but that true nature is usually beyond us, for every layer of experience reveals so many additional layers that we do not really know where to begin anew. What we do know is that we have never met an animal that has been reduced to a place in the same way some humans have been reduced to places.

The question came up late one night, late in one of those ever-increasingly philosophical discussions that heat up over the table in a bar in some small town after a long day's work in the field. Maybe Whitetail entered the discussion because of its beauty; after all, get a bunch of guys around a bar table late at night and the conversation often turns to beauty. The question was whether there are in fact any *places* in nature, whether in fact those places we call places might have innermost characters, innermost souls, and whether by simply peeling off the layers one might find that innermost nature that dictates not only our impressions of the place, but indeed our way of approaching and regarding and treating that place, as well as the place's way of approaching and regarding and treating the creatures around *it*. Late at night at that table *the* place was not even discussed. It was a foregone conclusion, since some around the table had been there before, that we would approach Lady Whitetail. The discussion centered around the excuse. We had been studying the relations between animals and the relations of animals to their environments and the relations of *relationships* to their environments, and Whitetail offers all of these things. A good excuse. We would chase *Semotilus atromaculatus* up Whitetail, for this pioneer fish participates in many relationships. In turn we would consider Whitetail an organism, and its meanderings through McGinley property an intestine. I arose bleary-eyed the next day to seek permission.

Both the headwaters and the mouth of Whitetail Creek are on McGinley property, which starts across the river from where we lived. Out on the Keystone road we had seen a burrowing owl and stopped. The owl drifted immediately to its nest hole with another owl's head sticking out. There were no prairie dogs in the village, the ghost town, only the two burrowing owls. The owls were on McGinley property, and sometimes in the evening we'd sit out in back of the lab and look a mile or two across the river and see the fields where the owls were. There would be people working in those fields, cutting hay. McGinley people. We were newcomers, strangers, and invaders in this land, and as such listened wide-eyed to stories of locals. Stories of locals had a way of making the McGinley property a no-man's land, of making the McGinleys themselves a formidable group. I had visions of ranch barons with guns, "men," cattle, kingdoms, a fortress of grass and barbed wire that might well be impregnable for a grown man looking for *Semotilus*. The grown man put his topographic maps in the car and headed for the McGinley ranch, straight up the driveway; asked directions to the man in charge; saw the sympathetic smiles of those ranch hands in the rearview mirror as they perused his out-of-town license tag, straight up to the McGinley house; rang the doorbell and asked for McGinley. The locals' stories were wrong. As the grown man drove back out the gravel driveway to the Keystone road, he had the feeling that this particular McGinley may have known almost everything there was to know about the *Semotilus* in Whitetail Creek.

Whitetail meanders off the McGinley property, then back on,

before dumping into the North Platte River. The creek is no sooner off the McGinley Ranch than it is diverted, used for irrigation, provided with pump houses, pipes, places where turtles lay, places where the railroad goes. There is no evidence from what we have seen of Whitetail that the McGinleys have ever done any of these things. On McGinley land Whitetail is wild, twisting, definitely not domesticated, brilliantly clear, triple cold, robust; with chiseled banks in places, a few wooden bridges for the Sandhills trails, and one place down by the mouth where some cattle and deer have waded across. It is the headwaters, however, that brings one back to Whitetail, for there is no spot in Keith County as beautiful as the Whitetail headwaters. Someone told us about the headwaters that first year, and we went up there. We stood on the bluffs and looked and took pictures, all in silence, for from up on the bluffs the headwaters of Whitetail Creek produce a sense of awe. As I write this I feel a credibility problem developing—how does one describe the scene from the bluffs to an easterner, how does one say to that easterner that the Whitetail headwaters rank among the most beautiful sights in the world, how does one say to a tourist from the Grand Canyon that up on the McGinley property the headwaters of Whitetail are one of the most awe-inspiring sights of all the earth? One simply says it, for it is true, and one simply says it with the conviction that if a person doesn't believe what you've said about the Lady Whitetail, then the person can go to hell. Or to Whitetail. I used to show people the slides of Whitetail headwaters, taken that first time, but for some reason these pictures were not very instructive. The beauty always preempted the biology.

The second year, with two vans of nonbelievers, we struggled through McGinleys' ranchlands again to the headwaters of Whitetail. Three out of twenty knew what to expect, and Crazy Diane was one of the three. Seventeen stood and watched, unknowing and slightly startled, beside the trucks. One stood and shook his head, for he knew what the race was all about. Two raced for the edge of those bluffs from which one looks into Whitetail's headwaters, and Crazy Diane won that race across McGinley's pasture. It was an all-out footrace. The two stood on the rim for a couple of minutes, panting, before Crazy Diane plunged into the headwaters pit. Minutes later, a hundred or two hundred feet below she was a single tiny figure, swatting flies, when the Frisbee was thrown.

It was the Frisbee throw of all Frisbee throws, gliding out over the headwaters for eternity before slowly turning to the north, past the sand bluffs, to be joined by the great blue heron. The heron and Frisbee flew together for a quarter of a mile, aerial formation duet. The "Skater's Waltz" welled from the rivulets of the headwaters bog before the Frisbee clattered to a halt in the buffalo berry bushes. The heron was long gone.

However, I am getting ahead of the tale; but that is understandable, since it is indeed difficult to resist the temptation to race into the innermost layer of Whitetail, the headwaters. We had decided to study Whitetail simply because of its beauty. This year I would seek the permission to approach Lady Whitetail, and *I* would choose the spots. I stood properly clad in jeans, denim shirt, lace-up boots, and Jacques seed cap on the gravel of Terry McGinley's driveway. The topographic map was spread on the hood of an old green Ford, and Mr. McGinley traced with a weathered finger the meanderings and property lines of Whitetail Creek. We talked of the place where it dumps into the North Platte River, of ways to get there. We talked of people who worked in town, who might not be able to give permission until later that evening. He understood completely that twenty people wanted to study Whitetail. He was not at all like another, an irrigation farmer, who thought the twenty were looking for a place to have a picnic. What Mr. McGinley failed to tell us was that our map-selected sites to encounter Whitetail, at least those on McGinley property, were among the wildest places in the world. What McGinley failed to tell us, as we traced our goings on paper, was that our goings would lead us to places where you were alone, although with twenty; where maybe nobody had gone for ten or thirty years; where there was not enough time in the life of the universe to see Whitetail. One discovers pretty quickly what McGinley failed to tell us when one parks on the Roscoe road and walks into the place where Whitetail

[95]

empties into the North Platte. It is a wild, wild place. There is a heavy wooden bridge, on McGinley property, and the McGinley fence is in perfect repair, but the mouth of Whitetail is a wild spot. Later, we would search for some other wild spots in Lady Whitetail and find instead the domestic.

I had visualized Whitetail Creek as a fifteen-mile-long intestine, and we were studying parasitic worms, those white creatures that so often have the movements, in a pan of water, that elicit fear, fright, loathing . . . but love in a parasitologist! One would like to think that the environment of the headwaters of a fifteen-mile-long creek would be different from the environment of the mouth, that the habitats afforded along the way would provide niches for creatures not found in other places, that a parasitic worm that lives in the intestine of a fish but requires, for the completion of its life cycle, the presence of several kinds of other animals might be found only in some parts of Whitetail. The exercise was a study of the requirements of associations among nonrelated species of living things.

Semotilus atromaculatus, the ubiquitous minnow of plains rivers and streams in this part of the world, participates in many of these kinds of associations. Furthermore, *Semotilus* is a pioneer, an opportunist, and no pioneer was ever afraid of a lady, no matter what her many moods, so the spotted tail minnow would then be our quarry as well as our excuse for what has since become known as the Whitetail Creek exercise. *Semotilus* can be chased with ease, but the associations of *Semotilus* require another order of magnitude of understanding. The fish is a coterie of friends, when approached as an example of a participant in a relationship, for not only is the special relationship with Whitetail there in that spot on the tail, not only does the minnow do better than many if not all other creatures at maintaining a relationship with the lady in all but her innermost moods, but this fish of all fish writhing in a prairie river seine-haul also participates in the most complex and unreal of relationships: that with the worm *Allocreadium.*

Allocreadium is little more than a worm, often found in the intestine of *Semotilus atromaculatus,* but it spends parts of its life in other environments, especially other animals. We mentioned this kind of life earlier when we spoke of *Physa* in the South Platte and the effects of that snail's style on the life of a parasitic worm. We mentioned a complicated textbook question, something about genes and gene expression and sets of chemical reactions necessary for genes controlling the development of a worm to be "turned on." We should also have mentioned that this course of worm events is so common among species that live in other species that one can usually assume even an unknown worm lives a life *somewhat* close to the average, the typical, the textbook generalization. Not being the resident expert on worms, therefore, it should not seem strange that my first impression of the life of *Allocreadium*, found in adolescent condition in the intestine of *Semotilus*, was that the animal *probably* (1) grew to maturity in the minnow intestine; (2) laid some eggs, which were shed from the intestine; (3) required that a snail eat those worm eggs before the eggs would hatch; (4) lived several larval generations inside the snail before emerging; (5) *might* have stayed inside the snail instead of emerging, and stayed there until a snail-eating creature came along; (6) spent some time as a slightly more mature larval stage in a nonsnail animal, perhaps even a small fish, and (7) became an adult only after a certain required time within the intestine of some vertebrate animal whose internal environment released the worm genes coding for adult characters. Although these assumed likely events may seem like quite a trek through the ecosystem just for some worm to reach adulthood, such treks are nevertheless exceedingly common among trematode worms, and trematode worms are in turn exceedingly common among familiar animals. We would do the Whitetail Creek exercise to determine if *Semotilus* was able to follow the Lady Whitetail through her wild and domestic times and places and in the process would hope to discover whether *Allocreadium* could tag along. A nice exercise. It would have been nicer had it worked, but then not all lab exercises work even back in the city.

A lab exercise that fails to work in the city is a true botch, but a Whitetail Creek exercise can fail miserably without a person's knowing or minding until weeks or perhaps months later. The beauty is there to distract attention from failure, the incredible beauty of the headwaters. The cutting edge of the lady's innermost personality is also there to offset the beauty. Who would have thought at the time of this famous exercise that a fairly extensive understanding of *Allocreadium* might well be required before a fairly extensive understanding of why *Allocreadium* is found in *Semotilus* could be developed, and that a fairly extensive understanding of why *Allocreadium* is found in *Semotilus* might well be required before the exercise to establish the distribution of *Allocreadium* and *Semotilus* among the Lady Whitetail's many moods could even be started? Hell. Even an advanced textbook hint about the life of *Allocreadium* would have been the smart thing to acquire before approaching Whitetail. That textbook hint would have told us that *Allocreadium* probably didn't live in snails first at all, but fingernail clams, so we would have looked for fingernail clams. The textbook hint would have told us that *Allocreadium* could also grow to maturity in beetles and caddis flies, not requiring a vertebrate, so we would have looked into some of the trillions of caddis flies we found at site number three.

I have been roundly criticized for not acquiring these textbook hints beforehand. Now that the Whitetail Creek exercise is a thing of the past, now that my plans for next year's Whitetail Creek exercise are well formulated, I am able to put this criticism into some kind of perspective. Let's just say it this way: if I decided to approach some incredibly beautiful lady with the idea of uncovering her many moods, of peeling layer upon layer off that personality, of finally getting into the innermost part of her being—the part that she originates from, and that ultimately dictates her style regardless of how domestic she might be—then I would probably not go to some textbook for hints. If I did, I would not admit it, and if I did and it were discovered, I would probably deny it unless faced with proof! Sometimes textbook hints have a way of shutting your eyes to things that should be seen whether they seem to be of immediate value or not in understanding the problem. Sometimes it's a better experience in the long run to simply blunder into a problem. There were twenty who blundered into the Lady Whitetail problem, but their blundering was not physical, for by the time they started they had become skilled at the dissection of creeks.

We had spent almost five weeks learning how to do this, the Whitetail Creek exercise. There is nothing quite like a seine to convert a person into a "can do" kind of person, regardless of how much a "does know" kind of person he or she is. The white pan works almost as well as a seine. There was no doubt about the roles of the teams, the jobs to be done, the animals to be collected, none whatsoever. After five weeks these people, twenty in all, did not barge into Whitetail; they stepped softly, each careful of the places others might require; they looked carefully and seined gently; they panned gently, they did not throw things, they handled the sodden log as they would, years later, handle a newborn baby. They sat on the banks with the deer tracks and thought, then wrote their thoughts. They cooperated, they called to others when an unusual sight was seen beneath the rotten bark of that submerged log, and the others came quickly but gently and appreciated the mayfly larvae. They held the barbed wire fences for one another. They had no idea the headwaters would be waiting for them the next afternoon, but even that knowledge would probably not have alleviated the feelings of helplessness that resulted when twenty young people and a guy in his second childhood started with the mouth of Lady Whitetail, at that place where she places that mouth upon the North Platte River, and discovered that the Lady Whitetail might not be so easy to dissect after all. It's lucky we didn't have many textbook hints to obscure the thing we actually discovered in the mouth.

There was really only one thing discovered at the mouth of Whitetail, namely, that it took more expertise than could collectively be found in the twenty to do the exercise. We collected mayflies and tried to collect snails; we made lists of birds we heard; we noted the tracks of wild mammal beasts great and small; we noted the turtles, the swallow's nests under the bridge, the water striders in the swirling eddy, the cottonwoods gnawed by beaver, the mosquito larvae in the isolated rain pool, the

Stagnicola in the backwater with the cattails, both probably brought from Keystone Lake; and we seined after *Semotilus*. Oh most assuredly *Semotilus atromaculatus* was there, in the mouth, and in the lab it was noted that *Allocreadium* was there in the *Semotilus*, but the critical animals to serve as checkpoints in an *Allocreadium* trek through the ecosystem were not there. We concluded that the fish brought the worm into Whitetail from elsewhere.

Then some turned to the mayflies, for it was the mayfly larvae collection, from the submerged logs of the stretch of creek between the McGinley bridge and the North Platte River, that finally emerged as the most significant observation of the day. The fish were dealt with easily, but the mayflies, no. Some concluded that we did not know enough about mayflies to do the Whitetail Creek exercise, and those some were the few who stayed in the lab through dinner trying to identify the first larvae. They were also the few who stayed in the lab through the volleyball game and were still in the lab as the rest left for town, maybe ten or eleven in the evening. It was only later, much later, that night, late late in the laboratory, late with the fire and microscopes, maybe two or three in the morning, dull with the beer of a visit to town, that the twenty reached a consensus. Regardless of the condition of their audience, the few were ready to report on mayflies, and held the rest of the twenty spellbound with the intricacies of mayfly identification. Near dawn the twenty agreed they did not know enough about mayflies to do the exercise. Incompetent. Approach a lady at a wild spot, find some of what you're looking for but get shown in return a bit of wildness you are unable to deal with, and you are ready to scuttle the whole affair. We probably should have called Terry McGinley. I sense he would have told us that if we couldn't understand a wild lady, then we ought to find a tamer one. Or at least a tamer spot in the wild lady.

There have been attempts, described in history and literature, by men frustrated with their inability to decipher the unknown in some bit of lady, to solve the problem by domesticating the wild beauty. We would see whether the domestication of Lady

Whitetail eased the discomfort of ignorance, an ignorance not easily shed, and we would find a domestic spot somewhere between the mouth and the headwaters. Place number two was up on the Wujek property, and Mr. Wujek had diverted Whitetail into a small pond for his irrigation pump. I had had this feeling, speaking with Mr. Wujek and his German shepherd about permission, that he was in for a treat when the twenty would arrive later that week. We met out on the road. He was driving his tractor, on his way to move some water pipe, and his dog followed. He was obviously interested in having someone to talk to for a while, and he was not about to allow me to go onto his property in search of a study site without him. He left his tractor in the road and got in my car. I cringed, for there was a pistol and a dead rabbit beneath the front seat, although the two were not particularly related. A parasitologist always picks up road-killed animals; some of my finest tapeworms have come from road kills. I really didn't think much about Mr. Wujek and the rabbit and pistol until his German shepherd climbed into the back seat. It was about two hundred yards back to the Wujek farm, and the dog almost found the rabbit. I was alert to dogs anyway that day, for I had earlier left McGinley property and, still seeking permission, stopped at another farmstead along the way. There was a small marsh on the map, and small marshes are always of interest. The marsh I could see from the car lay a quarter of a mile across a farmstead strewn with rusted machinery and car bodies. The railroad track crossed there, along a high fill, and the marsh lay beneath the fill. I viewed the scene from a locked car with windows rolled up. The sign as I approached the farmstead warned of the dog. The sign was correct. I backed off— the people worked in town during the day. We would not study this part of Whitetail. Mr. Wujek's dog, tracking down the dead rabbit in my car, was a delight. Mr. Wujek, learning that we did not really intend to have a picnic after all was also a delight and wanted to talk biology.

Returning with the twenty to the Wujek place, twisting through the overgrown homestead and back to the irrigation pump beside the cultivated field, we stood on the high banks of

Whitetail and looked out over a thoroughly domesticated lady. There were some items dumped into the creek, some additions to the landscape. There were this time also some snails, *Physa* pioneers, and *Semotilus*, and later *Allocreadium* in *Semotilus* again taken as perhaps false evidence that the fish came to Wujek's from the North Platte. The irrigation pond was continuous with the main creek through a barbed wire fence, and at once the hardy lads attacked the irrigation pond with a large seine while the women of the aquatic vertebrate team took to the creek itself. The irrigation pond turned out to be a part of every mud hole and long-standing prairie civilized tank, complete with large and unhappy snapping turtle. We don't really need to go into the animal list from that day on the Wujek farm, for it was a typical domestic list. It might have been compiled from a place in Oklahoma or Kansas.

I stood on the banks of Whitetail that day and mentally tallied the list as the twenty made their discoveries, and decided that this was after all a tame place. The pump house was on the high bank, a barbed wire fence had to be negotiated as one slid down the sandhill to the creek, and on the opposite shore the prairie was beaten down by cattle tracks. I stood facing the south as the creek curved toward me in a slight arch. To my left the girls were seining places where some large concrete slabs had been dumped into the creek and were tearing their seine. Across, sitting among the cattle tracks, a couple lit up a cigarette and contemplated their gallon jars of snails. It was about time to leave this place, and I counted people before wading into the creek just to get my jeans wet and pick a snail. To my right, upstream, the creek disappeared behind the high bank we had been standing on. It was a simple impulse, that idea to walk around the bank upstream, to wade the few yards up around to where we had not been. That impulse uncovered the wild lady again. It could not have been fifty feet around that bend, out of sight of the pump house, but Whitetail around that bend might as well have been on the McGinley property. Don't have to look too far beyond the domestic to find the wild spots in this one, I concluded. Back up on the bank, loading the vans, I surveyed Wujek's Whitetail with the knowledge that wildness is shallowly hidden just around the bend from domestication, and I concluded that Stanley Wujek might have purposefully domesticated the lady just enough to water his fields. And no more. That was a chauvinist interpretation; the pump and irrigation pond might actually have been about the maximum any human was capable of domesticating Whitetail Creek. In either case, Stanley Wujek came off pretty well.

It was only out on the road, heading for study place number three, having deposited Mr. Wujek back beside his tractor, that my thoughts returned to the headwaters and McGinley property. We were to be working upstream from the mouth, saving the best until the last. Site number three was about two miles downstream from the headwaters, and it was also very beautiful. It was also one of the most instructive hundred yards of creek and immediate surroundings I've ever waded into. It was a fine preview for the headwaters. Site number three would multiply the effects of the headwaters several times over, for site number three had none of the uncontrollable and startling attributes of the headwaters. We would discover at site three that the Lady Whitetail was at her best for company; charming, gracious, still wild but willing to give unusual experiences, tolerating our catholic friend *Semotilus*, teasing us with caddis flies she knew we didn't realize the significance of. The twenty responded in kind, and it was maybe the most memorable few hours of an endless summer to watch twenty students deal with the Lady Whitetail when the lady was at her best. That afternoon it would also be a memorable experience watching the twenty deal with bitch Whitetail at her worst, having been led to expect the best! I still have not decided about Whitetail at three, whether she really was sincerely at her best for us, or whether she was simply setting us up for her worst. In retrospect I refuse to give in and play her game; I simply don't care whether she set us up or was sincere; couldn't care less. Whether she intended to or not, she gave up something of substantial value at site three, her caddis flies, and I now have that something of value and she will not get it back; ever!

Whitetail at site three was very twisted, thrown into quick

folds, with many places where the banks were undercut, and the girls with seines went in more deeply than they had intended. There were snails at place three, *Physa* pioneers, small, out on the mud where cattle had stomped the bank into a flat few yards. There was much poison ivy, and there were many many barn swallows, nesting beneath the wooden bridge. There was an amazing channel or tributary of Whitetail that entered at place three, and following this channel, wading upstream, one became progressively more enclosed by the steeply rising higher and higher banks above and the increasingly deep water below. I stopped before the water got to my chest and before the poison ivy leaves over my head actually touched. And Whitetail was roaring out of this channel with a vengeance. There was not a place on the banks or sand of the nearby pasture where kangaroo rats and prairie voles had not left their trails, there were deer tracks everywhere, there was a long dead and well rotted snapping turtle, but most of all at site three were the caddis flies, and caddis flies take some thought.

There have been so many times when I've decided that if a student, a citizen, any person in any walk of life, would simply sit down and think seriously about a caddis fly, then that person would walk away from the experience with a new perspective on life and the world. That is a big charge for a small fly, but the fly is up to it; and if a single fly is not, then only a few different kinds together will be. Two major kinds occurred in Whitetail at site number three, and they occurred there in exceedingly large numbers. The fact of their occurrence there meant place three was different from the other collecting sites, if only because of the caddis flies, present only in rare numbers if at all at the wild mouth, and certainly not present on the domesticated Wujek farm, and very much absent from the headwaters. In the hardrushing cold of Whitetail three there were scattered branches, wild plum perhaps, or sumac, dead and black, wedged in the sand, and hairy with caddis fly larval cases. The cases hung evenly spaced, and as the branches were lifted from the water they swayed quickly, gave the appearance of moss, covering the twigs. The twenty's initial reaction was that these cases were part of the branch itself, but further and closer examination set them back a notch. The number of these cases was beyond comprehension, more like the number of snails in Lake Keystone. Every bit of submerged wooden vegetation was coated with them; Christmas, it was, so many, so like icicles, a fringe on a sodden twig. The bridge pilings were similarly covered.

One has to visualize the life of these insects beneath the rushing-hard cold of Whitetail three: some kind of food was coming down that creek in large amounts, at a very rapid rate, and was being trapped by these larvae. My mind goes back to the branches beneath Whitetail three. There were not really that many, about one submerged limb every ten yards, but there were many twigs and lesser branches along the banks, dangling and submerged, also covered with caddis flies. The animals had obviously taken up all the available space on those twigs ranging from one the size of a pin to one the size of a railroad tie supporting the bridge. The fact of these flies' dependence on twigs for home-sites was impressive only until one looked at the larvae with a hand lens. Each larva lived in a house, constructed by itself. Each larva's house was to the untrained eye so similar as to be identical to every other larva's house, the detailed architecture and accoutrements of each house built according to the same set of blueprints, and each not only fastened to twigs but *made* of twigs. One sensed no colony of caddis flies, as one senses a colony of cliff swallows, but rather sensed a set of instructions within each fly larva that chose twigs to build a house, arranged and glued those twigs in an identifiable pattern, and finished the job by adding exactly two much longer twigs, so that the final house resembled a tube with runners. As if that were not enough, they came in all sizes, all built according to the same exact plan. My experience in this country is continually that of standing in some body of water up to the knees, thoughts preempted and taken over by what has just been seen, mind racing to figure a way to tell the world this exists, and better judgment reminding me of my basic ignorance of the very things I must tell about. I didn't know what kind of caddis fly builds this type of house, but I would go to the literature and find out, only to discover that my original thoughts about caddis flies were pretty simple and un-educated. Whitetail three is a pretty simple place for caddis fly larvae, and the things they do with twigs in Whitetail three is but a fraction of what they do with other building materials elsewhere.

The caddis fly story does have an end. Neither Whitetail nor the literature contains a species of caddis fly with the exclusive species character of building its house out of other caddis fly larval cases. There are those in the literature, however, that build their houses out of snail shells. No, they don't expropriate the shells, moving inside them like a hermit crab. They glue the shells together to form a tube, then live inside the tube. Archi-tects upon architects. Architects are appropriate companions for beautiful women at their best for company.

The headwaters came that afternoon. They are located in a depression, maybe a sink, that cuts the water-bearing sand and allows the underground river to well through the surfaces in thou-sands of hydrants. The depression is exceedingly green from on top the bluffs and is really nothing more than a spring covering about a quarter of a section and set down below the surface a hundred feet, maybe two hundred feet. The walls of the head-waters depression are sheer sand in most places, although as Whitetail leaves its headwaters spring one can climb back out pretty easily. It is important to be able to get back out of White-tail headwaters, and often quickly, depending on the weather and your mood.

It is very easy to get into the headwaters bog: simply jump off the cliff. The sand is soft and one does not get hurt badly by falling, hitting the sand every thirty or fifty feet. If you'd seen a former football player do it, apparently without much prior thought, then you'd be willing to try. The ballplayer stood at the bottom that first year, laughing up at the rest of us, dusting the sand from his shirt, switching from dusting sand to brushing flies as he stopped laughing, still looking up. The girls joined in quickly, then it became a game. The second year, only Crazy Diane leaped from the bluffs first, and she was into the bog before the rest of us arrived at the rim, so her leap was not really observed. It didn't need to be; the headwaters pit is exactly the kind of thing any self-respecting biology student would leap into without thinking.

The view from the rim affects different people differently, but it never fails to stop someone for at least a few moments. There are different tones to the voices as a group of twenty sees the headwaters for the first time, there are excited pointings to different parts of the spring, and there are those who linger on the rim just as there are those who leap after only a minute's look. Today the great blue heron held people. Herons always hold people; people will always stop and watch a great blue heron. The heron would have been spectacular without the Frisbee.

The two Libras stayed on the rim, although one stepped a few feet down the cliff side, apart from the group, and the other sat on the sand and wrote and looked and thought. There was a place on the far side where a whole wall of sand and grass cover had slipped, and one Libra noted that and wrote it down. The cliffs were drilled with holes where animals lived, and the one Libra noted that and wrote it down. A naturalist talked about the vegetation as best he could, and the one Libra dutifully wrote it down. Another naturalist talked about the planaria in the spring hydrants, and the one Libra wrote it down, but the other tapped her foot in the sand and waited for the formalities to end. The one wanted to look and think but the other wanted to race. Both being Libras, there was little doubt that sometime in the future the one would want to race while the other would want to look and think.

The twenty (minus Crazy Diane, now visibly agitated so far below) were almost in a trance on the rim. One could sense that same feeling from the group that one senses immediately before simply taking something of great beauty, for the group knew that in a few short moments they would be in the bottom of Whitetail Creek headwaters. To be in the bottom of the headwaters pit is to have left that part of Keith County where total beauty hits you between the eyes and to have taken two steps forward and three gigantic thirty or forty foot steps downward to a point where the beauty surrounds and overwhelms, almost

Great Blue Heron

"Herons always hold people; people will always stop and watch a great blue heron."

too intensely, almost too much, almost so much that one has to
keep reminding oneself that this place is actually spectacular. The
decision to simply take something of great beauty is a very exciting decision, and there is always some fear of an exciting
decision, some desire to choose your own time for that decision,
which competes for your desire to carry out that decision.

The one Libra may have sent vibrations to the group, for they
waited. The second Libra sent some other vibrations to one in
the group, directly to one, and the one doesn't really remember
whether he was actually invited by word of mouth to race down,
or whether he responded to the vibrations by inviting the child
to race, or whether a tensing, a leaning, a give-a-way look, in the
body of the second Libra started it. The race lasted about four
seconds and the second Libra won. Actually the race lasted about
one and a half seconds, for the girl kept her feet when she hit
the sand cliff forty feet below, while the man lost his and went
tumbling the rest of the way. The world switched to slow motion,
and the man tumbled and tumbled and tumbled, and up on the
rim the group very slowly raised its hands.

The man sat at the bottom, in the sand, and the flies began
to gather. The girl asked if he was okay, and he laughed and
checked his camera, which had also gone tumbling for a hundred
feet through the sand and air. For months after the four seconds,
the four seconds would come back to the man, for various members of the group would find themselves in the comfortable social
situation and finally get up their courage. They needed courage
to ask if the man had been hurt by the fall. One by one they
asked over the next few months, Did you really get hurt? There
is still some sand in my camera, especially in the shutter speed
adjustment. There is still some Whitetail headwaters sand in my
boots, but then there will always be Whitetail headwaters sand
in my boots, even when there are no boots. At the time the fall
didn't seem to make that much of an impression on the group,
for the group followed exactly the same path and exactly the
same methods, but without the tumbling.

To jump into the Whitetail headwaters pit is to simply take
what you want, to simply take the beauty. The group came down

with high expectations, and the pit set them straight quickly. One does not simply take something of world-class beauty and dignity and performance; the things of that sort will always resist. Whitetail resisted, for the deer flies are a special part of Whitetail headwaters, and the headwaters pit does them better than any place in the world. The twenty stood startled at the bottom. The twenty immediately began looking for ways out. Weeks later, the Libra would be asked again to go into the headwaters and would refuse. One wonders, often, about the lives of famous and beautiful people, what it would be like to be married to one of those incredibly beautiful and worldly women. Why it is they seem to be able to supply the media with their difficulties, why Michaelangelo sculptures can carry the undertone of devastating frustration for the artist, why a van Gogh in our museum must be viewed with the knowledge of a lost ear and a mental institution, why a Beethoven concerto is heard within the context of deafness. One finds out why and what it's like, sitting in the sand dusting off the camera at the bottom of Whitetail Creek headwaters cliffs. The lesson is never forgotten, and it comes with the flies. Take Whitetail's innermost beauty, mister, and you also get Whitetail's innermost deer flies.

The genus is *Chrysops* and even they, in slow motion, would be beautiful. Their posture is total confidence. They are a precision machine, quick, purposeful, and exact, with none of the batting about and bumbling stupidity of a housefly. It is their posture that earns respect, and their posture would earn respect even if they were unable to bite. But bite they do, and through clothing and in the middle of the back and in large numbers; and their bites are very very painful. The twenty are trying very hard to concentrate now, and as they move through the other planet world of Whitetail headwaters they bat and slap continuously; some are already moving toward higher ground. The seining team searches among the rivulets for enough water to support a fish, and finds a land full of water but no fish.

Not even *Semotilus* has made it into the headwaters; this ubiquitous chub has stopped short of the innermost mood of Lady Whitetail, although we do not know whether it was forced to or whether it made the decision on its own. The failure to collect *Semotilus* should tell us something about the headwaters, but then that something is really nothing we didn't learn more quickly from *Chrysops*. That something is that the Lady Whitetail is at once a world-class beauty and a world-class bitch up there in the middle of McGinley property!

The headwaters is a spongy bog, and as the twenty move over the quarter section they do so in knee-deep mud, but one steps into a hydrant and is up to his armpits in a flash. It is very silent in the pit; there are two red-winged blackbirds and that's it. Later someone would find a baby duck, and we would all ask why there should be a duckling and no adult. Later someone would also find the remains of a duck up on the rim, and I would put some feathers in my shirt for a feather quiz—ducks are easy, sometimes. In the pit, the flies slash through cloth and skin: razor slashes, cuts, cuts on the arm that hurt all the way down your legs. I wince as I write this. Concentration is lost. The twenty tolerate the flies for a time, for the pit is science fiction, truly another planet, and there are plants there, and planaria, and physical beauty, a texture to the land, an arrangement of common things, that no one has ever imagined, much less seen. They laugh at first, then become serious. The flies are serious business. Whitetail Creek is Pandora's box. The twenty move faster at their assigned tasks, then move up the sides of the rim downstream. Some are moving faster than others, and they find the poison ivy. Some barge through the poison ivy, somehow thinking poison ivy is better than flies. Some find their way around, walking a mile to get a hundred feet up, away from the pit.

It is later in the afternoon now, reassembled at the vans. There is an air of beer in the group, and a not-so-subtle suggestion is made of a stop at Sportman's Complex. Not a person lingers on the rim. There is no talk of the beauty of the headwaters, at least now. There is a desire to leave, there is no identifiable reluctance to leave. The group is ready. We talk of how it might feel to be a cow in Africa, in tsetse country, and the group understands.

The decision to return to the mouth of Whitetail is a scientific decision, as is the decision to return to Wujek's and to site three. It is embarrassing now, to think of not being adequately prepared, intellectually, for those three places. The homework will be done next year, some name will be on those mayfly larvae and caddis fly larvae before we wade knee-deep into those places. We will have the records of last year's exercise to study, and those records will be studied. The methods of preparing for class are well known, tried and true. The information and observations will flow next time from those three places. There will be tension in the air as we approach the headwaters from downstream, for the physical nature of Whitetail and the animals and plants of Whitetail will change as we go from the mouth to Wujek's to site three. Someone in the next group will have been to the headwaters, and will talk of it in terms the rest cannot understand. The rest will understand on the second afternoon, however, and understand well. The decision to return to Whitetail's headwaters is a very different decision than the one to return to Wujek's.

In the months that follow that experience with the Lady Whitetail, that experience into which I so blithely led twenty second-year people, I sit at times late in the evening with Whitetail pictures, or early in the morning with a book on caddis flies, and at these times I think about the other women we met and dealt with in Keith County. The Ackley Valley Ranch is the girl next door; Arthur Bay is your high-school girl friend, the first real one; Keystone Lake is your sister's friend; Big Mac is your sister; and the South Platte River is the town putout. The Lady Whitetail is a New York model, and what you see on the cover of *Vogue* is but a fraction of what she really is; what you learn from looking is but a fraction of what you learn from participating. Robert Ruark wrote about these same kinds of things, but his tale was of people rather than places. This is a tale of places, but places can grow to be people and people can be reduced to only places. Robert Ruark's tale is called *The Honey Badger*. We will return to the headwaters next year; there is no way to stay away.

It is the middle of the Nebraska winter as I write this, and the middle of the Nebraska winter is a time to reflect on the past and future, a time to put into a framework the events of a life, so that when the spring thaw arrives one is ready and equipped for new lessons from nature. My son, John III, is playing on the floor, playing cars. He is six this year, and he has become an appealing child, his teacher says, a handsome, sensitive, perceptive, vocal, and confident child, an artist and an engineer, a reader and an architect. There will come a time, I know, when society will demand that he also become grown, that I take him aside in this ritual and teach him what there is to learn about women. This knowledge brings a feisty and mildly diabolical smile to the face of the father: I expect that when that time for the manhood ritual arrives I might just put little John in the car and drive to the McGinley property where just he and I will redo the Whitetail Creek exercise.

9

The Fundulus *Chronicles*

K EITH COUNTY has a way of inundating the human with
      nature, as you may have figured out from what has gone
before, and the human finds himself turning first this way and
then that as the attractions parade across the senses. The wrens
compete effectively for attention, but one step into the canyons
and the thrasher's call erases the wren, only to be erased in turn
when the human drives out the gate past a cliff swallow colony.
The snails compete effectively for attention so long as there is no
caddis fly in the pan, and the caddis fly in turn competes effectively for attention so long as it has no protozoa upon its back
or wedged into the leg joints. They cannot all be grasped; they
cannot all be studied. There may be some unwritten rule that
says a biologist is forbidden to know everything about every
creature, and that rule may be what follows from the ever-enlarg-
ing interface between known and unknown that is generated by
the study of even a single one of these inhabitants of the Sandhills. It is at once a sad and a happy choice for a man in muddy
jeans, when that man makes a decision to pursue only one of
these creatures, for a sense of happy relief is coupled with the
knowledge that only so much time remains and a decision to follow the killifish shortens the list of others to follow.

   One would like to think this decision was made from a basis
of careful thought, careful consideration of the best animal to
pursue, only after elimination of less scientifically "productive"

options, but one cannot in any truth admit to such a practice. The choice to study the little striped fish was nothing more than an impulse, and a fairly emotional impulse at that. Out of all the things in Keith County, the biologist chose for his own the plains killifish, and he chose it on impulse at that. On the other hand such a choice was not made out of ignorance, for early on the killifish was shown to be not a fish but a community, and the community itself encompasses both the artist and the laborer, both the classic and the romantic, both the ballerina and the boor. It has turned out to be a good choice. Other biologists, including one notorious Stephen A. Knight, student, agree and commit their time and talents to the killifish, and the tale of this commitment has come to be known as the *Fundulus* chronicles.

Fundulus kansae, the plains killifish, should by all rights be restricted to the Smoky Hill River in Kansas. Any fish with the name plains killifish and any fish with the specific epithet *kansae* should be found only in the Smoky Hill. The external appearances of the two simply go together in a way no other, more familiar kinds of animals seem to go together with their environment, and furthermore, the only real scientific study of the natural history of *Fundulus kansae* was done in the Smoky Hill. The Smoky Hill is crossed south of Abilene as one proceeds out of Nebraska along Kansas State Highway 15. Having seined the plains killifish, it is then difficult to actually drive a car over the bridge. At a minimum, attention is distracted from the oncoming traffic. At a maximum, someone slams on the brakes, the station wagon fishtails across the highway, coming to rest just off the gravel apron, and the driver is out shoveling through the back for a seine long before he realizes that this is just a family trip and there are children and toys in the back of that station wagon, not seines, microscopes, and buckets with portable aerators!

I have never set foot in the Smoky Hill, but there is no way the river can go wrong in my eyes, because the knowledge that *F. kansae* does its thing in the Smoky Hill means that in my mind the river can only be a wild, mysterious, spectacular, intellectual oasis on the prairie: a home, a refuge, as well as the most prominent feature on the map between Lancaster County, Nebraska, and Houston, Texas. This is the kind of feeling, the attitude, that pursuit of an animal species can produce in a person exposed even ever so briefly to an environment immortalized in the scientific literature as the classical home of that animal. Although I have never set foot in the Smoky Hill, I am totally convinced that when all of life becomes too much a hassle, the bills pile up, the committee reports are due, the research grant has been rejected, the automobile needs major repairs, the wife has a headache, the children grown tall and sullen, the dog crapping in the yard and the sod webworm eating it, then one only has to make the short drive into Kansas, step into the river with shoes on, and all things will become immediately rectified. They may not be rectified, but at least they will be placed into some kind of perspective. This rectification-perspective-wade-in test also works in the South Platte River beneath the I-80 exit ramp at Ogallala, for the plains killifish is not, after all, restricted to the Smoky Hill.

Fundulus kansae tends to be caught in the same seine-haul as minnows, and thus finds itself sold as "tiger minnows" in the bait shops around Big Mac. No one raises the plains killifish commercially, so tiger minnows in a bait shop tank reveal instantly that someone has been seining the South Platte for commercial quantities of bait minnows. Tiger minnows in bait tanks are usually floaters. The tiger minnow is not much of a bait minnow, for it apparently does not have the domestic options of a real minnow. The tiger minnow is not even minnowlike; it is not in the same family as minnows but is nevertheless caught in the same seine-haul as minnows, out of the same ten-foot-wide shallow channel of the July South Platte. The single haul may net five hundred fish. Four hundred ninety will be minnows, two may be adult *F. kansae*, and in July the other eight may be first-year killifish. There are *Fundulus* and there are minnows. There are no game fish to be seined in the South Platte.

The *Fundulus* chronicles soon demand a person become intimately acquainted with the business end of a seine. A seine is two poles, some netting, a float line, and a lead line at the bottom. A "keep the lead line down" award is given annually. If the lead line is not kept down, *Fundulus kansae* swims not only under and over the seine, but with its speed, around the seine, around the seiners, and it may, if the occasion demands, bury itself in the sand and gravel and simply allow the lead line to pass overhead. The fish may also hide up under the algal mats that begin to appear in the South Platte in midsummer. The uninitiated seine upstream, but the river, although small, is too swift for two humans to catch *Fundulus* by seining upriver dragging a twenty-foot seine. It is evidently classier and reveals less ignorance if one seines downstream, but then one has to run through knee-deep water not only fast enough to stay ahead of the current, but also fast enough to stay ahead of the fish. Sometimes the two speeds are not equal. In the South Platte it is not difficult to keep the lead line down, except in holes around pilings where a sudden chest-deep ice water bath may take one's immediate attention from the line.

One can also seine across stream, which in July tends to be easiest and most productive. One can also employ stompers, or recruit volunteer stompers, which multiplies the catch, particularly of *Fundulus*, many times over. Two humans walk very quietly along the South Platte sand, the stompers follow at a respectful distance, the seine is stretched, lead line down, poised at the edge of the water, and at the word "go" the seiners hard charge into the river, poking the sand with their poles, while the stompers hard charge a little faster and stir up the sand, with *Fundulus* ahead of the seine. The haul is quickly perused on the opposite shore. Minnows are thrown back toward the river, but they land short only to flip continuously on the sand. Fry wash out through the net and partway back to the water. Humans kneel in the sand and seriously attack the haul. *Fundulus* go in one bucket with a portable aerator; everything else is thrown back, and those that are not thrown back are washed out when the seine is flipped. Only rarely is a single minnow of particular beauty examined carefully. The humans are single-minded. There are minnows and there are killifish. If there were game fish, then I suspect there would be minnows and game fish on one hand, and *Fundulus* on the other.

The only way the minnows can get a moment's attention is to be so much more beautiful than their fellows that they stand out in a seine-haul of four to five hundred writhing little fish. Being *Fundulus* gets consideration and care. Being a minnow gets being thrown back *at* the water. Being an outstandingly beautiful minnow gets a minute's attention, and normally enough of a toss to actually get back into the river. The beautiful minnows truly are beautiful, and generally perfect specimens in breeding colors. The *Fundulus* are not beautiful except to the scientist. In fact they are highly variable in color, gnarled, and as a group beat to hell physically.

Their gnarled appearance does not come from mistreatment in the seine; it probably comes from being beat to hell as a child. The immatures—the fry, the first year fish—are uniformly pretty, although variable in color. Older fish are gnarled and often slightly twisted. Their heads are most variable in shape. A three-inch *Fundulus* reflects in its head two or three years of being beaten to hell by the South Platte River, raging at times, bone dry at others, and of burrowing into the sand and gravel, presumably to avoid predators. Later, in the aquarium, one can distinguish individuals by their head lumps, and their shapes give them a character a group of minnows would not have.

There are a number of places where the South Platte can be seined for *F. kansae*, the most popular being Brule, Ogallala, and Paxton. The word *Brule* has never failed to bring a particular look to the faces of people, although the look itself cannot be described. It's a half smile, which sometimes has a slight raised eyebrow, sometimes involves only the eyes. The look usually says "I've been to Brule." There is nothing unusual about Brule; it's simply a small town out west of Ogallala. There is a bridge and the river is very accessible at Brule. There is no Ramada Inn, however, and so Brule does give the impression of being slightly more pristine than Ogallala. While there may be no Ramada Inn right on the river, there is nevertheless a commercial sand and

gravel pit on the river. Edward Curtis also visited the Brulé—
the Brulé Sioux, that is—with his camera, and anyone who has
carefully studied the Curtis pictures of Brulé Sioux, especially
after having been to Brule, trades his "I've been to Brule" look
for a slightly different "I've been to Brulé" look. The latter look
is a look of pride, of a satisfaction of actually having stepped into
history for a moment through Curtis's photographs and emerged
on the South Platte River a few miles west of Ogallala.

It is now impossible to seine Brule without constantly looking
over one's shoulder. There is a brave on a horse watching, an
"I've been to Brule" smile on his face. The brave knows the killifish. He dismounts and wades across the river to the seining party,
leading his horse, which stops to drink. The brave glances back
and waits for his horse, then approaches the seining party. All
heads turn toward the brave, and there are some exchanged glances
of apprehension among the college students. The brave squats
beside the seine, and his eyes quickly survey the catch. Deftly, he
plucks out the killifish and tosses them into the bucket with the
portable aerator. He sifts through the rest of the writhing catch
and plucks out a *Notropis lutrensis* in breeding colors. The *Notropis* is thrown into midstream. The brave washes his hand in
the river and nods at the seine; he's signaled he's through, throw
the rest back, the raccoons will pick up the ones that don't make
it. He looks in your bucket with the portable aerator and smiles
with satisfaction that you've also kept only the killifish. He's gone
now, on his pony, and a brown dog follows. Sitting cross-legged
in the willows is an old, old woman, her blankets almost covering
her head. The blankets are very ragged, and her face is very
wrinkled. "His name is Little Striped-Fish," she says. "He is
different from the other braves."

For the students, Brule has just been added to the list of spots
that will pass the rectification-perspective-wade-in test, a performance that is of course dependent not only on that special relationship between beast and environment, but also on that between
man and beast. I should say it is especially dependent upon the
relationship between man and beast, for in the man-beast-environ-
ment connection that allows a spot to pass the test, the beast-

environment part of the chain is one that was forged eons ago and remains strong. Thus a man often subconsciously chooses a beast that will strengthen the weakest link in this triparty relationship. The zoologist in the man, however, knows that the relationship is within the beast and that there are few species indeed whose personalities can be bent to form the relationship a man desires. The zoologist in the man knows there are animals that quickly become domesticated when brought into the house. The zoologist in the man also knows that *Fundulus kansae* is not one of these.

For example, *Lepomis cyanellus*, the green sunfish, might as well be a puppy. One may sulk for a day; and if there are two, and the aquarium has no hiding places, and the two are of very different sizes, they may play games with one another for another day. But sunfish are voracious eaters and learn quickly the source of their food. The animals may not actually become so domestic; their feeding rituals may simply give the impression of domestication. Nevertheless some kind of an understanding between human and fish is very very quickly established in the case of *L. cyanellus*. The sunfish will eat crickets, roaches, flies, frozen liver, frozen spleen, and earthworms, as well as an assorted array of smaller fish. The speed of their attack is frightening. Furthermore, within a very few days the fish will come to the side of the aquarium and stare hungrily at any human who comes to the side of the aquarium and stares curiously. They will respond to the raised hand, although I've never seen a green sunfish actually strike at the idea that it might get some food. If fed on any schedule from any established feature of the room, such as the refrigerator, they will bounce around the aquarium, yes indeed like a puppy, when the master simply goes to the refrigerator. Later, they may do the same thing when the master enters the room. *Lepomis cyanellus* is either very dumb or very smart. It is difficult not to attribute the puppy-dog behavior to learning, in which case the fish is very smart. Dumb or smart, the fish is able to train the human.

The human responds very quickly to the behavior of the sunfish, and he is eager to establish a feeding schedule, so that this bit of wildness known as "sunfish striking a cricket" can be incorporated into his daily dull life. The human, normally within two weeks, which is really a pretty short time for a human, has *also* learned that he can make the fish all excited simply by going to the refrigerator, so the human goes to the refrigerator regularly and watches the fish out of the side of his eye. In about another two weeks the human learns that the fish gets excited simply when he comes into the room, so the human smiles when he comes into the room, he smiles a smile of satisfaction and pleasure, and supported ego, that some wild creature has learned to recognize him and gets excited when he comes into the room. It has been a long time since someone or something has gotten excited simply when the human came into the room. In any case, it is doubtful the fish knows how long it has been, and there is little doubt the human would be just as easily trained if it had been only five minutes. The human is either very dumb or very smart. It is difficult not to attribute the growing human affection and the added fish-feeding ritual to learning, in which case the human is very smart.

To date, *Fundulus kansae* has demonstrated no ability to train or be trained, with one exception. They will learn what a dip net is for and will frantically burrow into the gravel when a person enters the room with a dip net. There is no way they can know the fate of a netted fish, unless they can see to the lab bench, and it is difficult even then to believe they could make the connection between being netted and being killed.

Fundulus sciadicus does not normally share the South Platte with *F. kansae*, although the two are obviously closely related and can both be seined in an afternoon's work in Keith County. *Fundulus sciadicus*, the topminnow, lives in Cedar Creek and some other spring-fed tributaries of the North Platte River. The topminnow is not a minnow any more than the plains killifish is; however, *F. sciadicus* is very skillful at training humans, and it makes a fine aquarium fish, really better than *L. cyanellus*, since the topminnow will eat commercial tetra/guppy food. In nature *F. sciadicus* spends most of its time at the top of the water, rising in a miniature version of a larger game fish to strike at surface food. In nature, *F. kansae* spends most of its time picking insects off the rocks and gravel and burrowing into the gravel when a shadow passes. *Fundulus sciadicus* relies on speed; they are exceedingly fast and the human must run much faster, and usually through deeper water, to catch them compared to the effort needed to catch *F. kansae.*

One wonders if there is any connection between the fact that the topminnow is a surface phenomenon and the fact that it adapts so well to "captivity." There is a biological connection, of course, in that an animal that feeds normally at the surface will be most likely to strike at floating tetra food. But one wonders if there is a nonbiological connection between the ability to form a new relationship with a very strange creature, a human, and the fact of life at the surface. One also wonders if a burrowing life also establishes in the mind of a killifish the ability only to perceive potential death. One wonders if living at the edge of an environment enables a creature to establish relationships that are beyond those that can be established by a creature that retreats into the innermost parts of its environment. Relatives sometimes have a way of putting your best friend's characteristics into a different perspective.

We have chosen to follow the killifish, those of us who live the *Fundulus* chronicles, and now we find that the killifish has relatives that are also to be included, relatives whose traits put those of the plains killifish into a perspective for us. We have chosen to follow the killifish out of impulse and emotion, but not out of ignorance; we have found our places where we are in tune with the vibrations of the little striped-fish; we have mastered the seine, sometimes; we have explored the relationships we have come to expect from our friends and their relatives, and we have explored even more deeply into the basis for those relationships between men and fish. As we said before, the killifish early on has been shown to be a community, a community within the larger community of Keith County. Just as we have chosen the killifish out of Keith County, we have also begun to sort through the killifish to find, among all those organisms that make up a killifish, the one with enough of a life-style to sustain the chronicles.

Now with the introduction of cousin *sciadicus* we find we must ask whether the cousin is also a community, for the zoologist in the man has certain uses for this fish, this topminnow, if its community relationships are different from those of the killifish. The topminnow's failure to retreat into its environment has caught our eye, and its ability to survive the trauma of domestication means to us that possibly a living bit of Keith County can be brought to the capital city where a chapter of the *Fundulus* chronicles might be written in the glazed winter. Schemes of scientists. The *Fundulus* chase now becomes serious business, and the pursuit of these fish is escalated. We will go to Cedar Creek, where lives that cousin that refuses to retreat, and we will bring that creature into our lives as business.

Cedar Creek is not the kind of environment one retreats into; there is something about Cedar Creek, and Sand Creek and Sandy Creek and Whitetail Creek and Lonergan Creek and Clear Creek, that invites exploration. The creeks are very similar in that they are all spring fed, their headwaters consisting of those places where the aquifer is clipped by the terrain. Their waters are cold, flowing even in midsummer drought, and crystal clear. They may be diverted locally to provide a pond for irrigation or they may be allowed to ramble, as in the case of Cedar Creek, until some overgrown child with a seine finds the place they cross the road. There is not much to be seined in these places: no bass, no trout, no walleye; only suckers, and minnows, such as *Semotilus atromaculatus*, and upon occasion, *Fundulus sciadicus*. The seining of *sciadicus* differs markedly from the seining of *kansae*, and it is less a seine than a chase. The instrument is a "six-footer" rather than a "twenty-footer," and it is applied to Cedar Creek with a spectacular abandon. The activity is conducted at a run, for example, at a place where Cedar Creek flattens into a series of loops, soft sand, poison ivy, and deer flies. It is serious business, but it looks like a joke. It is no joke to *F. sciadicus*; they will be used as experimental animals in just the same way as *F. kansae* are.

Humans, considering those the world over, do an amazing variety of things for sport and for living. The daily activities of people on the far side of the globe may seem so exotic to us as to be in fact, in our minds, unreal. To a tribesman in the New Guinea highlands, the sight of campers bumper to bumper on Interstate 80 for three hundred miles may seem so exotic as to be in fact, in his mind, unreal. If one calculated the amount of human energy devoted to activities universally considered productive and normal—for example, fixing up the house, working as a plumber or electrician, doing an unknown job in some office— and compared it to the amount of human energy devoted, all around the world, to activities considered frivolous, only entertaining, or outside the inner-sensed universal definition of "normal," one would find the former amount to be but a small fraction of the latter. Humanity does not really work, nor does it like to work, nor is "work" in one society the equivalent of "work" in another society. Indeed, work in one society may be the wildest form of play in another. The work of seining *Fundulus sciadicus* must be considered the wildest form of play to the social subset known as normal Americans. It would be very, very difficult to convince the average heavyset woman on an Omaha street that the seining of *F. sciadicus* was a normal and required and serious part of the business of being a biologist. It would be far easier to convince a Sandhills rancher: his life is so tied to the land that he understands in a moment the efforts of any person to get close to even a minute part of the Sandhills known as *F. sciadicus*. Besides, the life of a rancher includes a lot of activities well outside the realm of reality for the average American, so the rancher has something in common with the seining party. To drive over the hills, off the road, at fifty miles per hour in a pickup truck and count cows is a normal and expected part of the rancher's life. On the other hand, it's the kind of thing that a citizen of downtown Los Angeles might contemplate doing for entertainment. So the rancher standing on the road north of Roscoe understands the seining party. After all, the rancher is at least somewhat alone and paranoid: neither the state nor the nation understands his economic plight. Alone and somewhat paranoid, he empathizes with the full-grown professor, also alone except for a couple of students and somewhat paranoid about being a professor, in the middle of Cedar Creek. Besides, on this particular day both the rancher and the professor have on seed caps. Seed caps in this year are red, white, and blue, and some of their panels are mesh. These caps say "Jacques Seeds" on a patch on the front. "Jacques" is a French word, pronounced "Jake's" on the North Platte.

On this day the *sciadicus* are thrown into a bucket with a portable aerator. An attempt is made to seine Cedar Creek south of the road, but the attempt fails. Cedar Creek is too brushy, too twisting, too narrow south of the road. The deer flies are intense, and they bite through clothing, sometimes through more than one layer of clothing. Even in the bucket *Fundulus sciadicus* are beautiful, they are not beaten to hell, and in the aquarium they are even more beautiful. Back in the city you might pay two dollars for a fish like that. Why, here we pick them up for nothing, only had to drive fifty miles with four guys, two seines, three buckets with aerators, and spend two hours running through Cedar Creek where it crosses the Roscoe road. The fish are not striped, but the males in breeding colors have red fins and tails. They feed the first time. *Fundulus kansae* must also feed, since they do stay alive as long as *F. sciadicus* in the aquarium, but they must do it so subtly that the human gets no satisfaction. *Fundulus sciadicus* feeds the first time. Within a day their fear is gone and they respond strongly to a hand lifted as if to scatter guppy food over the top of the water.

The *Fundulus* chronicles now demand that the fish be converted into gills under a microscope, for upon the gills live the artist and the laborer, the classic and the romantic, the ballerina and the boor, the community that started our chase of the killifish in the first place. The gill of a fish is a thing of rare beauty, especially when it is the gill of a small fish and it is viewed under the microscope. Gills consist of four arches on either side of the head. The arches are known as bars or gill bars. In a ten-pound channel cat the gill bar is made of a tough rod of cartilage supporting the gill filaments themselves. In progressively smaller fish, until one gets down to the first-year *Fundulus*, which is an inch long, the cartilage supports are less stout. Thus one needs a pair of very heavy scissors or bone shears to cut out a gill bar of a ten-pound catfish. With an inch-long *Fundulus*, a gill bar can be removed with a pair of forceps. The gill filaments themselves are flat and arranged radially and stacked, like leaves, along the posterior outside curve of the bar. The filaments are really folds. The microscope reveals that the filaments have additional folds, the secondary lamellae. One can see through the gill tissue of a small *Fundulus*, can see the individual cells that form the gills themselves, can see the blood vessels within the filaments and the blood cells within the vessels. Blood cells are stacked within the vessels, and the vessels branch and loop.

It is not difficult, looking through a microscope at a *Fundulus* gill, to put oneself upon that gill, to become so miniature as to be a part of the gill filament world. In this dream you are wading thigh-deep in a bouncing, slippery, moonscape pulsing with limber tubing beneath your feet, the secondary lamellae quiver and flop over your shoes, and as far as you can see there are only folds and more folds and white cytoplasm and red blood. It is an eerie place, this microscopic world of a fish gill. It would be far less eerie if you were not sharing it with some creatures straight out of fantasy and science fiction. They are real, these creatures, they are a part of the community of things that live upon small fishes' gills, and they are guaranteed to boggle your mind.

Members of the genus *Trichodina* are gill surface symbionts and are made of a single cell; since they are of a single cell and hence not divided, they are sometimes described as being without cells. The question of whether *Trichodina* is cellular or acellular is exactly the kind of word game that we often use to distract ourselves from the significant and beautiful. We think, when we argue such questions and evaluate the evidence in support of either view, that we are participating in high intellectual activity, in high science, that we have made the transition from studying the hard facts of cold "original" science to applying those facts philosophically, conceptually, generally, and broadly. Nature and *Trichodina* simply could not care less who resolves the cellular-acellular question. In considering this question the human has ignored the beauty of the denticle ring. There are few things like denticle rings in all the world of life.

Trichodina is built like a small flat bell—a fringed bell, that is, for there are skirts of membranes and cilia that beat continuously and rotate the animal as well as give it linear motion. Thus *Trichodina* on a *Fundulus* gill looks and acts like a minature vacuum cleaner. When seen for the first time by any human, *Trichodina* rarely fails to elicit the same response: they look like beautiful little vacuum cleaners! It is the denticle ring, however, that eventually gets to the human. The denticle ring is inside the animal and lies flat, parallel to the undersurface of the bell and parallel to the gill surface. The human doesn't know what to do with the denticle ring; it has no satisfyingly obvious function. A satisfyingly obvious function does wonders for us when trying to decipher the structure of a living thing, for we like very much for our animals to have defined roles. Our desire extends to their parts, and the parts of animals are most comfortable when they have defined roles. Tiger teeth. Elephant ears. Zebra stripes. Denticle rings. For us the denticle rings have no defined role. They are there and all species of *Trichodina* have them. It would confuse the picture to add that the rings vary considerably in their structural details, depending upon the species of *Trichodina*. The limits of confusion might be reached if it were pointed out that many kinds of fish have their own particular kinds of *Trichodina* but that some species of fish have more than one kind. One immediately wonders if denticle rings are anything like wedding rings: many kinds of fish really acquire only one type during their lives, while other *kinds* of fish regularly acquire several. One could also wonder whether a collection of denticle rings would reveal something of the travels and experiences of a fish, as might a collection of wedding rings.

The basic part of the ring is a circular stack of cones. Visualize a stack of fifty ice cream cones, pointed ones like they had in the old days. Now lay the stack along a table and bend it into a ring, keeping the cones stacked. What you have is the basic part of a denticle ring. Inside *Trichodina* each cone has a spine, usually a long and delicately curved spine, that extends toward the center of the circle. Inside *Trichodina* each cone also has an outwardly radiating spine, also delicately curved. As if this were not enough, there is another ring made of individual parts looking for all the world like truncate French curves, but stacked their flat sides to one another. There must be several hundred of these plates in the ring that lies within *Trichodina*'s body but above the denticle ring. There is no obvious function for any of these structures. There is no obvious reason why *Trichodina* would have to have these things to survive on a *Fundulus* gill .

The protozoa and the rings are there, that's all; they are part of the fish, for the fish is not really a fish at all but a community. In the fish books biologists separate the fish from this community. Nature doesn't accede to the books, however, for there is evidently something about the fish gill that has elicited the artistic in the protozoan. Our use of the protozoan literature is straightforward enough: we go to the books in order to identify the protozoan. Furthermore, when we find new species of the genus *Trichodina*, we turn to the denticle ring for structural characteristics to describe the cell. Our use of the ring is a practical one, but it is our impression that our use of the ring is far removed indeed from the use of that ring by the cell-animal. We find now that in our chase of the killifish we have turned up a ballerina and an artist: *Trichodina*'s movements never fail to elicit an exclamation of beauty, a sense of awe and wonder in a student at the microscope as the one-celled animal glides and pirouettes over the secondary lamellae. Of course the ballerina in *Trichodina* is well known, but the artist in the animal, the sculptor of the ring, is not so well known. We cannot but decide that until "further research" debases the artist and explains the ring, that we are in the same position with *Trichodina* that we are with Picasso; none of us really knows why the creature builds the ring, although we make use of the ring, and the "why" in the animal is about as easy to extract as the "why" in the artist.

Beside the ballerina, on the gill filament, is the boor, as tied biochemically to the fish gill as the artist; and the romantic artistry of *Trichodina* is offset by the classic efficiency of the monogene worms. While the protozoan is a thing of almost nonbiological beauty, and it is the beauty that captures one's attention rather than the animal, a monogene worm is a thing of total practicality.

The monogene addresses the question of the fish gill quite directly: there is a structure of overkill efficiency at its posterior end, and this structure has a large double-hooked anchor, hooks and spines on fingers that are in continuous digging motion, and perhaps several suckers. The combined anchors, hooks, and suckers are the haptor, and with the haptor the monogene maintains its spot on the *Fundulus* gill. There are no problems here with satisfyingly obvious functions. What the protozoan accomplishes with mind-boggling beauty, the worm accomplishes with terrifying practicality.

The *Fundulus* chronicles are not ultimately a tale of artists and laborers, however, and in finally admitting to themselves that they are so captivated by the killifish community for scientific reasons, the scientists feel the slightest twinge of fickleness. Their habits are *somewhat* stereotyped. After all, they are scientists, and these habits say that a scientist, any scientist, goes to the ballet to see the ballerina for entertainment, goes to the museum to see the artist for feelings, and drinks beer at the local tavern with a laborer, but in the final analysis seriously studies an animal. On the other hand, even scientists regularly go back to the ballet, the museum, and the local tavern, so I suppose it is simply our great fortune that in pursuit of an animal through a community, we have turned up entertainment, feelings, and companionship to go along with the serious business of small fish.

The tale of a several-year fish chase is a tale of time and money also, and in telling such a tale a person finds himself stopping now and then to reflect on the accomplishments of the pursuit, to codify in one's own mind what has been derived from the activity. We have yet to mention the animal, the single species of animal whose life-style has so captivated us that we chose from all the organisms of Keith County this one animal to study, and yet in our enactment of the *Fundulus* chronicles it is this one animal, also a participant in the gill community, that gives to us all that any wild animal could hope to give a biologist. With the gifts, however, comes some teasing, for this animal is not about to give up easily some of its most secret secrets. The animal is *Myxosoma funduli,* another protozoan that lives within the gill

[129]

tissues of *Fundulus*. At first we attacked this animal from a vulnerable direction, one that really did not exist in the scientific literature, and the animal lost that skirmish and gave up some very unusual and interesting information. On guard now, the creature stiffens its defenses, and our attack becomes one of asking questions about the best point to probe: should it be biochemical, should we ask specifically and directly about the chemical relationships between cells, or should we go back to the field? Simply in order to discover where we are as humans trying to unravel the life of a community, trying to justify the time and money of a three-hundred-mile trip to seine little striped fish, we need to ask what it is that we have seen of the world through a one-quarter-inch seine, even before we consider *M. funduli*.

There are some things we have, those of us who live the *Fundulus* chronicles, that others do not have, but our sense of possession is not a selfish one. The fact that we have these things, the fact that only a fish chase has given them to us, is a fact that drives us to tell others what there is to be gained from pursuit of nature. We have seined *Fundulus kansae* and *Fundulus sciadicus*, and we have read in the literature of little striped fish in the Smoky Hill River, and now we understand why a site can preempt our thoughts and direct them to the animal, just as the animal can preempt our thoughts and direct them back to some place in space or time when we met that same animal before. We are not afraid to give in to the rectification-perspective-wade-in test; and indeed we come to anticipate it. We have learned to seine, learned to extract our animals from their environments, and in so doing we have met the ranchers and we have thought seriously not only about heavyset women on Omaha streets but also about work and play and the methods for turning work into play and play into work. We have been to Brule, and because of that we are able to see things in the Curtis photographs that others cannot see; we have met Little Striped-Fish and the old woman in the willows, and we see them, old friends, in the Curtis pictures. We have taken stock of our relations with wild things in captivity, and have come to understand that sometimes it is the wild thing that is brought into the laboratory, but often it is the human that is taken into the wild by a wild thing that is brought home. We now wonder if we will allow ourselves to be conditioned by a fish again, but I suspect the answer to that question will be "yes", and maybe a "yes, on purpose". We have chased the two species of *Fundulus* into their homes and have seen the perspective that relatives can provide. We have put the gill of a small fish under a microscope, have seen the community upon it and within it, have decided that larger (edible) fish also have these communities, and have concluded that when a fisherman cuts off the head and throws it into the trash, he is probably throwing away the best part of the fish! We have found an artist and a laborer upon the gill, and because both are so far removed from us, we are able to see *in* them the artist and laborer; and we now try to see the artists and laborers, ballerinas and boors, classics and romantics in our human friends, and because we do we feel we are able to fight off the dehumanizing conformity pressed upon us by today's world of pride and money and power and poverty. Yes, it has turned out to be a good choice, the killifish.

We need to tell of *Myxosoma funduli* before the story ends, however, for *M. funduli* lives the lives available to a creature at all levels—from that of two cells trading molecules to that of two states trading the water of two separate prairie rivers. At certain times of the year, the *Fundulus* gill is swollen, especially at the tips of the filaments, with an infection of *Myxosoma funduli*. As is the case with many animals that live in close association with other animals, no one knows how the association is formed, on a regular basis, in nature. The biologist says, "The life cycle is unknown." We can also say that we simply do not know what events occur in the lives of these animals to bring them together. It really is a little sad to think about, but then on the other hand it is almost a commonly accepted thing among biologists. What is known about *M. funduli*, however, is enough to tantalize the imagination, for what *Trichodina* displays as formed beauty, *M. funduli* displays as a beauty of process. *Myxosoma funduli* forms cysts within the filaments of *Fundulus kansae* gills, and the walls of the cysts are made of membranes that drink the fluid from between the fish gill cells. *Myxosoma funduli* must thrive on this fluid, since within the cyst are hundreds of cells, each undergoing a series of changes that brings them into association with one another. Once the cells are in association with one another, their roles change, they assume structures they would not otherwise, and in some cases their whole nature is transformed so by the association and their assumed role in it that they are no longer recognizable as cells. The process is called spore formation, and in the South Platte in June it occurs many millions of times, with such regularity and according to such a specific set of instructions that the spores, once formed, are virtually identical in shape and have virtually identical measurements. The spores are spewed into the South Platte, and there our knowledge ends.

We know where the spores come from, for their formation can be seen with the electron microscope, and the cellular role-playing dance can be analyzed and ordered through the study of electron microscope pictures. We also know that within one of these spores there is a single amoeba, ready to receive the environmental cue that will unleash its genes and allow it to explode into another cyst upon another fish. Since the amoeba is a single cell within its spore, the equivalent of a single propagule in an elaborate space capsule searching for a hospitable planet, we know that when that amoeba becomes a cyst full of cells it will in essence be a compartment containing a clone. The tale of South Platte River scientific fact now approaches very closely some tale of galactic scientific fiction, not because of the place and time of these events; oh no, rather because of the issues and implications for all of society raised by the events of *Myxosoma funduli* sporogenesis.

The spore is formed in the following way: two cells within the cyst come in contact, and the touching of their membranes precipitates an assumption of roles by the two cells. The two cells are members of a clone—they could not be more equal genetically, yet their association results in an assumption of roles in which the activities of one and all its progeny are directed by the other. One of these cells envelopes the other. The enveloped cell begins to divide until there are ten cells within the one, and the ten then arrange themselves into two groups of five. This description of cells and their biology might just as well be a description of a folk dance, a seating arrangement for a first-grade reading group. When the two groups of five have been formed, then within each group the five cells further arrange themselves: two flattened ones on the outside encompassing one at one end of the developing spore and two at the other end. Having accomplished this final arrangement, four of the five die and leave the infective amoeba in its space capsule to be spewed into the South Platte River.

One last comment: the four do not die before accomplishing their task of building the separate parts of the spore, the valvelike shell, the capsules containing coiled filaments, and so on. This set of role-assumption events does not occur outside the first enveloping cell, *but it does occur within a clone.* Sitting at my desk, flipping through a large stack of electron micrographs, my mind moves quickly to some science fiction literature. Somewhere in that literature, as well as in more serious books, such as Toffler's *Future Shock,* is a discussion of the cloning of humans and the dangers thereof: a clone of incredibly beautiful women; a clone of unquestioning laborers; a clone of superior warriors; a clone of superintellects; a clone of anything. *Myxosoma funduli* seems to be telling me not to worry, that it doesn't really take a long time or much effort to reduce a clone to a set of role occupants, the functional equivalent of a "normal" society. I make some half-hearted attempt to resist the more extreme message from *M. funduli*: that there will always be some enveloping cell, some leader, to direct the activities of genetic equals and that that leader will arise when the genetic equals come together and exchange molecules. I make less of an effort to resist the idea, however, that role assumptions, such as those required for *M. funduli* to launch an expedition in search of a new fish, may also be required for any clone to go exploring.

The electron micrographs tell me about the lives of cells and suggest molecular relationships between those cells, molecular relationships that enable individual cells to assume roles, which may indeed force roles upon individual cells, and I must conclude that this cellular role-playing within the cyst within the gill within the fish *Fundulus kansae* is the set of events that dictates the concentration of spores in the South Platte River. The spores are out there in that river now. One can always see them in the mind from the exit ramp bridge although they are oh so microscopic. One knows that in nature, nature is functioning perfectly, and if not perfectly, then relatively perfectly compared to its ability to function when domesticated and brought into the lab. While I conclude that a study of the biochemistry of sporogenesis might well be of incredible interest within the four walls of my laboratory, I must, after living in Keith County, also conclude that in nature these biochemical events may play a minor role in the life of *Myxosoma funduli*. I sense that *M. funduli* carries out these events routinely, as a matter of course. I also sense that *M. funduli* carries out another set of events, that of finding another fish, in a manner that is not so matter-of-course.

The possibilities become endless. The spore may require months of interaction with its environment, the South Platte, before it is able to germinate. The spore may be destined to a withering death unless the fish performs some act, some act of fish behavior, that brings fish and spore together. Now this last one is an attractive idea, the idea of continued contact between spore and fish, all played against a backdrop of the South Platte annual cycle. Yes, this last one is indeed an attractive idea, for the Public Power and Irrigation District has made an unknowing attempt to separate fish from spore, and in the process given me a three-hundred-mile long laboratory to use. Not only that, but there appears in the newspapers some official's idea of transbasin diversion of prairie river water, from the South Platte to the Republican, perhaps, or to Little Blue. Our study of the killifish leads us, all in the same week, from the darkened electron microscope room to the river to the newspaper to the Irrigation District offices to state economics. Quite a bit to learn, I conclude early in the morning, from studying a protozoan that lives in the gill filaments of a little fish.

The spores themselves are now out there in the river, no doubt, caught up in the debris and sand of the South Platte, millions and billions of spores mixed up in the algal mats, reading the annual signs of the river, awaiting the passing fish, processing the environmental cues that say it is all right to leave the protective space capsule and initiate another clone, another set of role playing. The river's homeostatic but evolutionary history becomes manifest, in the minds of the scientists, in the biology of *Myxosoma funduli* spores. The South Platte River becomes not a river, not a bridge with a cliff swallow colony, not Mr. and Mrs. Retired Couple on the exit ramp, not the driftwood, the willows and annual weeds on the sand bar, not the water, but *M. funduli*. There is a link, a communication, between the South Platte and the spore, and what we read as weather reports for a whole year, as water flows, as temperatures, at the reporting station Ogallala, the spore is reading as a message from earth to cell. Maybe, the state has messed with that message, I conclude, looking at a map of the Public Power and Irrigation District developments. Downstream from Ogallala, at the city of North Platte, the North and South Platte Rivers converge, and at that point the rivers have been dammed, the water diverted into an irrigation canal, and the sediment dredged to be dumped on the banks in miniature mountains.

"What in the hell are those people doing with my spores?" The question is only a thought, but the student who shares the *Fundulus* chronicles is looking at the same map and thinking the same thought, and we know now we will soon be back to the field to find out just what those people are doing with our spores, to find out whether they are dumping our animal on the banks along with their dredgings, whether they are irrigating their corn with our spores. That we will find out. From the darkened electron microscope, with its plethora of reddened dials, responding to the touch of an experienced student in the same manner and in the same setting as a giant airliner responds to the touch of an experienced pilot, or as the Platte River responds to the touch of the Corps of Engineers, the killifish chase has filled our lives with excitement and thrills, mystery and a sense of accomplishment, a hobby, a love, a very feeling that we are able to see the importance of the river as well as the cells on the green screen of the electron microscope. Yes, we have learned a lot about nature from our study of the fish and its gill community, but it is obvious from the newspapers describing transbasin water diversion schemes that we have a long way to go.

After the twenty-first of December in this country, the days begin to lengthen, and by the middle of February there is a running channel through the Platte River ice. The frozen river is treacherous and uncertain to a shivering guy in waders, but it is rapidly becoming manageable with each passing day into March. We don't know where killifish is in December, when it's Arctic cold on the Platte and the exit bridge is slick with glaze; when the seines are dry, rolled in the top of the closet; when the aquaria are empty, the portable aerators silent, their batteries dead; when the buckets are stacked and dry, and students and professor sit in a city lab and stare the Ogallala blues at one another. Nor do we know where the killifish is in February or March, simply because we've not looked, but in February or March the days are longer, the sparrows are already sneaking grass into the martin house, the pear trees are already sneaking fluid into their buds, and the cliff swallows are somewhere down in Mexico getting ready to return to the exit bridge.

In February or March we will look for the killifish, for the ice is on its way out; and, furthermore, plans have been made for the summer, and those plans require the information that only a trip in February can provide. The summer plans include plans for the following summer. And the summer after that. And the summer after that. The summer plans made in December before a trip in February inspired by research results of the previous July do not include a plan to ever stop studying *Myxosoma funduli* and *Fundulus kansae* as they live in the South Platte River. Yes, it was a good choice to follow the killifish; the choice has cost us next to nothing, and it has given us next to everything. We have seen the world through a quarter-inch seine and have been impressed with what we've seen. The world has also seen us through a quarter-inch seine. We don't know whether the world has been

Belted Kingfisher

A fellow South Platte fisherman agrees, "Yes, it was a good choice to follow the killifish; the choice has cost us next to nothing and it has given us next to everything."

impressed with what it has seen, but then we don't really care.

The *Fundulus* chronicles in the mind now become a thing of boots, jeans, waders, seines, buckets, aerators, microscopes, slides and cover glasses, cold, hot, coffee, beer, country music, gasoline, station wagons, the exit ramp, sand—here's sand in your shoes—North Platte, Ogallala, the Smoky Hill, aquaria, electron microscopes, students and friends, student friends, stompers, Brule, Brulé, the old woman in the willows, Little Striped-Fish, Cedar Creek, relatives, work, play, Edward Curtis, the Irrigation District, state economics, denticle rings, space capsules, clones, roles, spores, dead batteries, learned journals, two-by-two slides in a projector, time, typewriters, plans for summer, the world through a quarter-inch seine, hooks and suckers, and artists, and Platte water in the Republican River. There are two kinds of Ogallala blues, the good kind and the bad kind. Both are good, but the bad kind is good only for the individual, while being trouble for a too-organized society. "Boots . . . (to the) . . . River" gives one the good kind, but it's still the Ogallala blues.

10

Painting Birds

THIS IS A CHAPTER about what it feels like to be taught. I have understated that; this is a chapter about what it feels like to be taught by the best in the world and then to try to do what you were taught. Maybe what this chapter is really all about is what it feels like to be taught, by the best in the world, to *try* to do something. The old man was the best in the world, and as a young man he had in turn been taught to try to do something by another old man who was the best in the world at the time. I call them both old men. I never knew one; he died in a car wreck before my time, but his pictures are still alive. The other I know would never be an old man if he lived to be 175 years.

"Doc" was always able to outrun and outwalk the younger people, he was always up earlier and stayed up later, always did his assignments even though he was not required to, always produced an unbelievable amount of writing, books, papers, was always available for interviews, was always a better shot with either a 12-gauge or a smooth-bore .22, was a teacher who gave a test every day and graded off for misspellings, commas out of place, and so on, and always had the papers back by the next class, was a scientist of world renown, and on top of it all could charm any matron into a stupor within a few seconds. He had done the latter a time or two, I suspect, in these later years, simply in order to avoid further use of his increasingly precious time by lay society. A matron affronted might well be a detriment to a beggar's career, but a matron charmed into a stupor first then politely excused was a different matter. We are all beggars in this profession. He could also walk through the woods and pick up a feather and identify not only its owner but the part of the body it came from. "Feather quizzes," we called them, down in some creek bottom in southern Oklahoma, when a rain-soaked and dried bit of rusty and buff feather was plucked out of the leaf litter and turned over and over in his hand, his mouth hardly able to hide the devil and his eyes going from student to student.

If one had just killed a rusty and buff brown creeper and looked carefully at the feathers, one might be able to identify that one. I know now, however, that if one had tried to paint a picture of a brown creeper, even years before, then not only would the owner of that feather have been obvious, but also the part of the body it came from and the pattern it produced before being lost. I have no idea how long the knowledge of a brown creeper's tertiary wing feather will last now that I have tried to paint the pattern in watercolors, but I know that individual feather's colors and pattern will not soon be forgotten. I cannot foresee a time when I will be unable to recognize that feather. I also think the old man must have painted a picture of every feather of every bird in the world.

He always used watercolors, and the fact that he always used watercolors now makes a watercolor somehow far more significant than an oil or an acrylic. I always wonder at the prices one can pay for any oil paintings when the old man painted in watercolor. Joseph Turner's watercolor sketches, hanging on the walls of the Museum of Modern Art, so completely overshadowed the finished oils that it was a shame they were in the same room. Of course, all this is only in my mind, and only because the old man painted birds in watercolors. As a youth he had been under the tutelage of Louis Agassiz Fuertes, and there is not another living person today who can really say that. The university was proud of his paintings, and they often hung in the library showcases for earthlings like us to study. The pictures were so very simple and yet so very alive. Every stroke was evident, at least to one who had tried, and the technique was so obviously simple and straightforward that even if one had never tried, one was ready to try after a few minutes' close-up study of a Sutton original. It never turned out to be quite so simple. The amount of anatomy contained in the borders between feather tracts was too staggering.

It is impossible to place a value on the things that happen to a person when that person by chance is touched by the life of an individual like the old man. Thinking back, the decision to become an ornithologist in March of one's senior year in college, two months away from a degree in mathematics and with no advanced biology credits on one's transcript, does not seem unusual. But then, it has been seventeen years since that decision was made, and George Sutton in retrospect made such a decision natural, easy, no-questions-asked. Truthfully, though, there were plenty of questions asked at the time, such as those of a sister-in-law who wondered through a clipped-out cartoon just what was an ornithologist in terms of dollars and cents. There were also questions in George Sutton's mind, and the fact that the erstwhile mathematician is not now an ornithologist reveals that he had asked the right questions of a student, like a matron, charmed into a stupor.

In my naiveté I had asked one day to see some of his originals. We made a date and I arrived at his home. I was poured a cup of his coffee: boiled with a bunch of grounds in a saucepan and served, also with a bunch of grounds, in a perfect china cup and saucer. He came into the dining room after a few minutes with a stack of paintings about a foot high, and I went through them one by one. There were some mistakes, some that had not turned out quite the way he'd intended. There were also some beauties, some shorebirds with complex patterns. There were some older pictures, with a background style he has since dropped. There were some newer pictures, done in the Arctic on textured paper. He patiently answered my questions. How does one make this pattern? The secret is to mix up enough paint, load the brush once, and after the pattern is dry, brush over it to make the feathers seem continuous. It does work, only it takes practice. It also takes the right brushes, a point he has made in his own writings. He had received his first set of right brushes from

Fuertes. I have had only one right brush, and the tip no longer points.

It was in my father's desk, and we found it after he died. There is no way to know how old it is, or what brand. It is simply black with a brass sleeve to hold the hairs, which are lighter than those of many brushes. I knew the first time it was a right brush. The picture turned out so much better than my perception of my own ability would allow. It was in the brush. Now that the tip is worn and it no longer points well, I realize it was in the brush. I priced some brushes and bought some brushes. A ten-dollar brush is not nearly the quality of the one found in the desk. The brush itself is not as large as a pencil, just a number 6 or 7 watercolor brush. There are some brushes this size in the bookstore and they cost twenty-six dollars. It takes a while to bring oneself to spend twenty-six dollars for a paint brush smaller than a pencil, especially if there is the lingering feeling that even a twenty-six-dollar brush will not be as good as the one found in the desk. My father always went first-class. I wonder where Sutton gets his brushes. I don't want to discover he could do the same paintings with a thirty-five-cent brush.

It was six years after the mathematician had decided to become an ornithologist, and the mathematician was now a parasitologist, studying a tropical disease, experimentally, in New Jersey. He had gotten to New Jersey by way of Kansas, where for three years he had lived in the Cheyenne Bottoms, studying bird malaria and painting. Still in his naiveté, he wrote Sutton and asked a personal favor: could he have a letter of introduction to someone at the American Museum so that he could get into the library and study the Fuertes originals? Sutton supplied the letter and the parasitologist sat at a long wooden table in the museum library, studying the Fuertes originals. There are many black and white photographs of Fuertes originals in the same drawer as the paintings. A black and white photo of a watercolor is very revealing: the medium is a value medium rather than a color medium. Remember that, mathematician-parasitologist, the next time you have a live brown creeper in one hand and a twenty-six-dollar paintbrush in the other. One cannot be associated with

Painting Birds

George Miksch Sutton and sit in the American Museum library with his letter of introduction, studying the Fuertes originals, and not try to become a bird painter.

The first thing one needs is a bird.

It was very very early in the morning and it was cold for midsummer. The weeds were soaking wet and the sand worked its way up into tennis shoes that were still filled with sand from the previous day's work. We waded only a few feet into the marsh before she came to the top of a cattail as they all do initially, scolding, fluffing. Ten feet away, she was dropped with a .22 pistol, loaded with birdshot and held Gene Hackman–style. The pencil sketch was done once, completely erased and done again, this time satisfactorily, since it began with the wing. I begin drawing with the wing whenever possible; somehow the rest of the body and its posture fall into place around a wing. The painting has been used over and over again. There are a lot of people who now know and respect the long-billed marsh wren who would have never thought about such a creature before.

Luckily she had not fallen into the water, and she lay on the dried stalks of last year's growth. There were a few breast feathers puffed out, twisted at odd angles, and they were straightened with forceps by preening. One has to try preening just like the old man, whenever one handles a dead bird. I have wondered so often about the act of preening, as any biologist must wonder about a ritual. A messed-up live bird will preen itself back into shape, an obviously functional pattern of behavior. I have watched George Sutton preen a dead bird and have said to myself, "This man has studied birds so long he has become one," since his very first act upon touching a dead bird was always to preen it back into shape. The bird always ended up in better form than it had been in alive, for the old man could preen a bird better than a bird could preen itself. He used, as far as any of us could tell, exactly the same motion as the bird would have used with its bill. None of us thought to check it out at the time, but I wonder now if he had several pairs of forceps, from which he picked depending upon the bill shape of the bird whose feathers he was about to preen. In retrospect all this preening was the ultimate respect for a living thing that had given its life for some human use. He had total respect for his subjects and succeeded well beyond the rest of us in his attempts to convert their lives into a set of values that would influence all our behaviors for the rest of our lives.

I stood in the marsh for a few minutes, preening the wren back into shape. It worked fairly well, and I slipped her into a paper cone. The old man always carried a supply of paper cones, and all of his students always carried a supply of paper cones. Never know when the guy driving in front of you is going to hit a kingbird, have to be prepared to pick it up, preen it, and cone it. She was done before breakfast. The colors came out of the right brush. There is something about going to breakfast having already painted a bird. It is my favorite picture. Later, I walked back and cut the cattail to go under her. The long-billed marsh wren is an animal version of a busted cattail. In the marsh a year later someone told someone else not to kill a marsh wren. I heard about it secondhand.

"Don't kill a marsh wren," he'd said.

"Why not?"

"They're his favorite birds."

A red-headed woodpecker is a different matter altogether. Particularly alive. Woody Woodpecker is a red-headed woodpecker. Woody Woodpecker is several orders of magnitude underportrayed. Maybe this was done on purpose, for Woody is indeed lovable. A cartoon about a real red-headed woodpecker might well be X-rated.

It was caught in the mist net strung between two hackberry trees down by the boat house. The mist net is Christmas several times a day; one never knows what kind of a treat is going to hit the net and become entangled, hanging upside down, feathers twisted, choking on loops of nylon thread, and feet and legs skinned and bleeding slightly from the tight twists. A bird hitting the net is supposed to stay alive, but unless one works quickly and gently, it may go into shock. There is a surprisingly good feeling one gets from freeing a bird from a mist net. The animal has suffered such indignity, its feather vanes are shredded, but in your hand it's still a wild thing to be studied, painted, held, helped out of the net, preened back into shape, stuck with a hypodermic needle to discover if it has malaria, and released. Yes, the feeling of releasing a bird after having painted it and sampled its blood is a very good one. The feeling one gets from preening a netted live bird back into shape is also a good one. It's easily done; with the right finger movements the seemingly torn wing and tail feathers suddenly are put back together. With several weeks of daily netting, the net becomes your life-focus; your obligation to nature, which is flying into it; your source of surprise and expectation. You hope that some new friend is in the net, you hope that it is not too entangled, you hope the bird will stay alive like it's supposed to, you hope it has malaria, you hope it's of the right age for your studies, but mostly you hope to all living hell that it is not a red-headed woodpecker. This day it was.

"We got a red-headed woodpecker," she said.

"That's nice."

"It's mean," she persisted.

"Oh?"

"It's in the heavy metal cage."

It's fall now, and I look at the painting done that day back in the summer with her red-headed woodpecker. It is the picture of a maniac. No painter, especially the painter of birds, would ever boast that he could portray the personality of his model in a picture. That is something that happens once in a while. Still, the field original of her red-headed woodpecker is a picture of a maniac. I don't show it to anyone. Now that the experience of actually holding that animal for two hours has passed, now that an eternal optimist's mind has screened out the stressful parts of that encounter, I feel an obligation to paint a little more gentle version of that picture. It's an obligation to the animal; I don't really, in October, believe that animal deserves to be portrayed the way it was in July. The picture has been done again, now, and I like it. It's not one of my best, but it is one of my favorites. How often does one get the opportunity to change the personality of a creature that presents itself as a maniac.

"I like that picture," I say, standing back, deciding if the values were right.

"It's not mean enough," she says. She is the same wife who had asked, back in July, when it was that the bird was going to be released. That was right before the thing went for my eyes as I was peering closely for detail. I have never taken a blood sample with such pleasure. I've also never been spit at by a bird, and in recalling the manner in which that woodpecker returned to the junipers, I have this impression that he spit at me as he flew away.

One was alive and one was dead. One had hit the mist net and one had hit a car. One was Western and one was Eastern. One was a living pair of scissors, and the other was limp and preened back into shape, the blood still draining from the mouth. One was shades of chalky olive and yellow, and one was shades of gray. One was all style and fortitude, and the other might also have been that if it had been able. One continually bit and talked; I don't mind a bird that bites or pecks unless, like the red-headed woodpecker, it has the equipment to do it effectively, but I love a bird that talks, tells you in no uncertain terms of the inconvenience of its present situation.

The western kingbird's vocalizations have about the same effect as the first rose, the first mosquito, an unexpected June vacation paid for by someone else. Any southern boy knows that when the kingbirds return, then tomorrow is summer. They used to nest in elm trees, trees of exactly the right size, not too large, not too small, just right for a kingbird. There were rows and rows of these trees along the streets of many Oklahoma towns, and the kingbirds were there the day before summer.

Nebraska is a northern clime for a southern boy. It is May in Nebraska, and the tenth of the month is that first frost-free

Red-Headed Woodpecker

"A cartoon about a real red-headed woodpecker might well be X-rated."

date. There is still condensation, sometimes firm, on your car windows in the morning. Kids are still wearing jackets, and on some days your breath still fogs. But now it is light early, and one morning you walk between the buildings toward the laboratory with only a few joggers passing, headed for Oak Lake, or the railroad tracks, or who knows where. You walk beside the stadium, which on fall Saturdays becomes the third largest city in the state, a citadel and monument to athletics, now very silent with dew on the Astroturf. You walk beside the old chemistry building, now the home of the journalism school, and you remember going into that building after the chemists had moved out. It was a science fiction nightmare; evidence of spills and explosions, chemical reactions climbing the walls, rotted pipes. You step from between the buildings and the kingbirds are there, chattering, taking over, asking you where you've been, it's time to get to work even though it is only 6:30 A.M. Okay mister, where are the bugs, quickly please, this is my tree, don't you know summer starts tomorrow? That's the morning you start with a tear. Even a grown man with bills and wife and teenage children and a mortgage and more cars than he can drive effectively can afford to start the morning with a tear if he knows it is the morning before summer. That's about the effect of a western kingbird's talk.

A month later, alive in the mist net, every impression of kingbird style is confirmed as true. Get me out of here you ass! And hurry, can't you see I'm tangled? God, what clumsy fingers! Easy on the feathers, pal! Jesus Christ, you pulled that one out! Who set up this thing, anyway, huh? The pleasure of having it in the hand, talking favorite talk, distracted my attention. The painting is terrible, one of my worst. I have this recurring dream in which I am scraping ice off a car windshield, it is five below zero, it is pitch black at eight o'clock in the morning, and the wind is blowing thirty miles an hour when a western kingbird appears. We've canceled winter this year, fathead, tomorrow is summer. Okay mister, where are the bugs?

Alive, eastern kingbirds are upright citizens in stylish gray suits, and the low telephone wires, barbed wire fences, and high weeds are their spots. Barreling along the interstate it is the barn swallow and cliff swallows, but turn off that interstate onto the gravel road and it is the eastern kingbird. Your car is watched carefully, but the insects it stirs up may be grabbed in an instant. Unlike the western cousin, which challenges a car on the Keystone road, playing dodge-'em for crippled dragonflies, the eastern kingbird usually watches. It may lift batlike in your car's dust and move on down the fence, still in your car's dust.

Today it was on its back on the gravel road beneath the towering sandy-earth dam. The dust of the car ahead was still in the air, and blood was coming from the mouth and one eye. It was preened and cleaned as best an amateur could do it. It is not a particularly good picture, the bill is not quite right; although that is not surprising, since the bills of flycatchers are very difficult to get right. Feather texture is also exceedingly difficult and requires years of experience. The Fuertes originals in the American Museum have perfect texture, and the different textures of the different parts of the body are obvious. For some reason the texture of the head and breast turned out correct, with the character of kingbird softness that one can see on an eastern kingbird even from the road. I have studied that picture many times, trying so hard to remember just exactly how it was that the texture was achieved, since it obviously did not happen on purpose. There is an interesting weed under the bird, and it is a weed I picked from the roadside where the bird was picked. It was a tall dried weed that somehow eventually found its way into a flower arrangement in an old milk can. The weed was fun to paint; parts of it are still on our kitchen table, sticking out of a vase our daughter made at a Girl Scout meeting from a Wishbone salad dressing bottle. The picture appeared in a magazine published by the state Game and Parks Commission and some lady wanted to buy it. I put an outlandish price on it. After all, it was the field original. She had just returned from Europe where she had been purchasing art and with her experience had decided the local boy could be haggled down to half the outlandish price. No way. One does not insult a dead eastern kingbird by allowing the price of a field original to be haggled down. I did offer to throw in the weeds, however. The picture might have looked nice in her living room, but it's also doing just fine in my closet.

Eastern Kingbird

"There is an interesting weed under the bird, and it is a weed I picked from the roadside where the bird was picked."

Painting Birds

There are fledglings who are considered sullen, but of course it's not a true picture. All one has to do is watch them with their friends, observed from afar perhaps, and they are not sullen. They usually talk well no matter what the occasion, often especially well if there is a complaint to be aired, and even do a pretty good job when it's their talk that has gotten them into a complaint position in the first place. They don't really allow adults to get their hands on them, even mentally, very often, and they have this very convinced feeling that they can handle the world even though there may be some evidence to the contrary. Their errors are usually errors of judgment rather than errors of intelligence, but so often they just luck out and squeak through, and then demand to be fed. There are also times when the adult simply feels the time has come for an adolescent to learn a lesson, and backs off to allow events to run their course after the kid has mouthed himself into a jam. Adult magpies careen up the canyons far in advance of humans anyway, even when humans have only insect nets, so it is probably stretching the point to think that this mother analyzed the people creeping up the road, answering her adolescent's complaints from the juniper stub, and decided they meant no harm, that it was time to careen up the canyon and let the adolescent mouth himself into a jam. Which he did.

It is a spectacular treat to get one's hands on a magpie, a live magpie! A red-headed woodpecker destroys the hand that holds it, and the assault hurts, and no feeling of love, only one of awe, wells up in the owner of a hand that holds a red-headed woodpecker. A magpie, even an adolescent, also destroys the hand that holds it, and the assault hurts, but you love it if it's a magpie. The world must love a scoundrel; there is no other explanation. It is a shame our lives are so dull that we must love a scoundrel; not that scoundrels don't merit the love, but it is still a shame the world is so dull. A world with a magpie in the hand is not dull. It's a world with a scoundrel to love.

The whole business was a Beatles movie joke: the bird was

in the middle of a twisted juniper high on the canyon side but along the road. Below the tree it was a fifty-foot drop to the water, above the road it was sheer walls of Brule and juniper another couple of hundred feet up, the parent was long since gone, and the adolescent was complaining, probably for food and attention. Only the Keystone cops go after a magpie with an insect net, but in a flash he was swept up with the same motion one uses on grasshoppers and cliff swallows. He complained the whole time, pecked with the strength of an adult, and sulked and sulked, drooping the wing in a most unflattering manner.

I smile as I write this; it was a trying two hours and nothing but pleasure came out of it. The picture is not that great. It reveals a sullen and complaining adolescent magpie, and people really don't like it very much. But special things about magpies are all that I see in the picture: the covered nostrils of all crows and jays, the white feathers that lap up over the shoulder from the breast, the special pattern of white and iridescent green on a single flight feather, and the battered tail. The white feathers are something special, and you have to hold the bird to appreciate them, their texture, their flair. The magpie in the hand is pure drama.

The bird is long gone now, flown up the canyon after the parent and also likely grown up itself after so many months. I get the same feeling, looking at the painting, as I do looking at my daughter's twelve-year-old birthday portraits. Shuffling through slides, looking for animals to show a class, and there is the box stuck away with the entire roll shot on her twelfth birthday. Some of her pictures are obviously of a sullen adolescent; but funny, all I see in them are the white feathers. Who knows how the kid's luck will run? As for the magpie, the parent's decision was correct; I could have been a coyote but wasn't.

They pass through your life in so very many ways; some are alive and feisty, some are alive and scared, and some are dead. Even on the dead ones you gently lift the eyelids to get the iris color. They fly into small places and get caught; they fly into glass places and get stunned and killed. After a few years you look at a mounted bird with a certain amount of disdain, the

Magpie

"A world with a magpie in the hand is . . . a world with a scoundrel to love."

feather tracts are never exactly the same as on an intact specimen. A mounted bird never really passed through anybody's life the way even a dead kingbird, hit by the car in front of you on the Kingsley Dam gravel road, did. There is never any experience that goes with a mounted bird. The painted ones are stuck in your memory forever, and they are things you have that no one else has and, quite possibly, no one else but a bird painter could ever have. There is the sick little blue heron, picked up from the leaf litter of a mixed colony; there is the skimmer chick done in the blowing sand off a New Jersey coastal town; there is the baby coot off a prairie marsh, the most untamed animal you have ever handled; there is the royal tern at Port Aransas, the rock wren from Cedar Point, the western grebe washed up on the shore of Big Mac, the fish crow brought in as a pet by your downstairs neighbor, the brown creeper, the Maryland yellow-throat stunned at the base of a city building, the junco lying dead on your window ledge; and the list goes on and on, and every item on it represents an experience that is branded on your brain and still burns its lessons and feelings and values there.

It is cold now, in the late fall, and in the late fall in this country a person cleans out the garage to make room for the car, to make room for the stack of dry firewood, to get up off the floor all the things that will be ruined by melting snow. There on a shelf is a broken cattail, and it falls down on the littered workbench. You don't even give it a second thought, but put the cattail back up on the shelf. It's been two years now, but it's the cattail you picked to go under the marsh wren.

11

Love, Joy, Money, and Pride

THIS IS A STORY about and for Steve Fretwell. It is also about people, particularly the manner in which people approach learning, research, teaching, and simply experiencing. It probably contains nothing more than has been written a thousand times before. The ideas here are basically those of Steve Fretwell, and indeed the very words, "love, joy, money, and pride," are Fretwell words.

You might be asking why a chapter like this when we are talking about western Nebraska, but the answer is very simple: I first met Steve on the hills above Keystone Lake. I hope he does not sue me for taking, incorporating, the things he said those days. When my students take things, incorporate them into their attitudes, thoughts, approaches, then I am proud. I feel satisfied, and during those couple of days I was a student of Steve Fretwell's. I am not necessarily one of his fans, nor am I a member of the cult that precipitated about him so readily. The guy just had quite a bit to say about love and joy and money and pride, and I now use those words every day, separating events and activities into their four categories. It helps, really helps, to understand what is happening to a person. If a person is doing something because of pride and money, when in fact society demands that the activity be conducted from a basis of love and joy, then something will be wrong. If a person decides to do something out of love and joy, but then society demands that he perform according to the rules of pride and money, then something will be wrong. On the other hand, if a person gets love and joy out of getting pride and money, then that person's life should take certain turns. For example, that person should never study *Fundulus kansae*, because there is no way in hell much pride and money will come your way from studying *F. kansae*. But, a person can get great amounts of love and joy out of the *search* for pride and money, and then *Fundulus kansae* is indeed his animal, for his search will never end, and by definition therefore, his source of love and joy will never end.

I didn't really meet Steve Fretwell on the hill above Keystone, at least not the first time, but I did *really* meet him, first, on that hill. We had some one-way correspondence: I was chairman of a committee to which ecologists were applying for work, and being an ecologist of sorts, Steve wrote many letters of recommendation for students and others applying to this committee. I thought at the time he must be some big shot, and in that same year was somewhat disappointed to discover that he was not really a big shot, as we normally define the term, but a guy just a few years younger than myself, just a normal, average college-professor-type guy. Except that he studied the dickcissel. Now there's a real queer, I thought at the time. I realize now what a big shot it takes to study the dickcissel. Jesus, I hope Steve doesn't sue me over stuff in this chapter; he must understand that these things have to be said, otherwise you won't understand the Ogallala blues. Worse, you may not understand why we go back to Arthur Bay and back to the South Platte River. Most of all, you may not understand why we think it is important to go back to Arthur Bay. My files are full of his correspondence for one candidate or another, but none of them were hired. He tried but it didn't work. Actually that line comes from the guy who was hired. My files are still full of Fretwell correspondence. I met him the second time at the projector in an ancient building on the University of Nebraska campus. I was sitting for no special reason by the projector, and Steve was a big shot guest lecturer that afternoon. People were gathered from offices and buildings around to hear the big shot. The big shot walked to the top of the auditorium to put his slides in the projector, and I introduced myself.

"Dr. Fretwell, I'm John Janovy; we've had some correspondence."

"Oh yes." (He was very busy trying to see if his slides were in order and turned correctly. They were neither.)

"I sincerely appreciate the informative letters you've placed in our files regarding the candidates from your institution."

"Oh? Yes. May I sit down here?" (He had a small bit of paper in his hand, maybe a three-by-five card, and he was drawing a graph. Only later did I realize that he was actually drawing the graph in preparation for the hour's speech *he was about to give*!) "Nice to meet you." I thought his seminar was terrible. After all, I had stumbled across the dickcissel every day in Oklahoma, and thought I knew it well. There must be twenty million creatures to study other than the dickcissel, I used to think, and in fact still think. What I know now is that someone must study all those twenty million, and some person who studies something like the dickcissel is probably going to think the study of something else is weird. No, that is incorrect. A person who studies the dickcissel, or who at least appreciates and understands *why* someone else studies the dickcissel, is not a person who will think the study of species X is weird. Just the opposite. The connection between love and joy and the dickcissel becomes pretty obvious after a very short time. If we are ever going to place the kind of value on our world that is required to preclude its total destruction as a planet, then we must be able to find the love and joy in a dickcissel, we must accept without thinking "weird" Steve Fretwell's trek to South America on the heels of the migrating dickcissels. Following the dickcissel to South America will have to become in our minds the *expected*, the logical action that stems from our set of values, rather than the queer. I suppose I was rather easily educated. After all, although I did think that following the dickcissel to South America was queer, I would have followed the yellowlegs just as far and farther, had I not been married and childed and mortgaged. But I see now those last three things are convenient excuses. Fretwell followed the dickcissel under the same circumstances. I left before his seminar was over. I had seen a dickcissel and a dickcissel nest, and I had heard a dickcissel before. He was a pretty poor speaker. I was not impressed with the fact that he prepared for his speech while sorting his slides. The dickcissel bit still lingered, though, and I was at least relieved to know we would have something to talk about in July, when I picked him up at the bus. We could talk about dickcissels all the way to Ogallala; it's only a six-hour drive. Needless to say, the next time I met Steve Fretwell was at the Trailways bus depot. He had on purple corduroy jeans.

I was returning to the South Platte River for a few days, returning with my eldest daughter, to seine *Fundulus kansae*. It was July, and I was returning to do the things I had not had time to finish earlier. It was my time, my trip with *Fundulus*, and my assistant who was already at Ogallala; it was my time to reflect on the research, on the directions to take now, studying nature, to gather the thoughts that would be of use months later, to file some thoughts for use maybe years later, to simply allow the electrical charge that builds when a person wades in the South Platte to settle in and become an actual part, incorporated, of oneself. That's what kind of a time it was, and was going to be. It was a week to seine little striped fish, to do the research with the veil of teaching pressures and organizational responsibilities lifted. It was also a convenient excuse to return to Ogallala, to take the chance that a short trip would alleviate the Ogallala blues, knowing full well that the trip might make them worse. There was another excuse, too, and that was Steve Fretwell's short course on quantitative ecology. I was then, and still am, a pretty dumb ecologist—a good excuse to get some concentrated learning from a top-notch guy while at the same time knocking down a few *Fundulus*. He was riding the bus to the capital city; I had agreed to drive him to Ogallala. The Sunday morning was spectacular. He was sitting on the sidewalk in his purple jeans reading a paperback. He got in the front seat, and Cindy moved to the back. The three of us talked for a while before the conversation turned to science in general and dickcissels in particular. The guy grows on a person. We, or rather I should say he, noted every dickcissel for the next three hundred miles. Cindy passed judgment at the first gas stop.

"Dad?"

"Yes, Cindy."

"What is a dickcissel?"

"It's a little ugly bird with yellow on the breast and some black on its throat."

"Why is this guy so hung up on dickcissels?"

"The dickcissel does things that he wants to learn about. I don't think he studies the bird for its beauty, because it doesn't have any. I think it just changes its way of life very drastically when it migrates, and I think he wants to learn about that."

"I think he's weird."

"He probably thinks people who make ceramic pots are weird, too."

I knew the minute I said it, it was not true. If there is one thing a person is able to do, if they've ever done *anything* out of love and joy, that one thing is recognize another person or another activity that proceeds out of love and joy. No person ever made a ceramic pot for pride and money, and if they did it probably netted them neither. The only artists who ever got pride and money out of their work were artists who worked because of love and joy.

"I doubt it," she said.

The lectures, conducted on the hills above Keystone Lake, were not quantitative ecology. Nor were the exercises, also conducted on the hills above Keystone. The lectures were on values and reasons for doing things, and the exercises were lessons in asking questions. It became very obvious very quickly that I did not know how to ask a question. It became just as obvious just as quickly that of all the things there are to learn of on this planet Earth, the vast vast majority, the absolutely overwhelming vast majority, require an approach characterized by love and joy. For example, there is little if any pride and money involved in the study of yucca. Furthermore, the study of yucca requires a patience derived only from a sense of love and joy. Sitting on a rock, being told by a person I now considered a good friend and spectacular teacher that I could not ask a question, sitting on a rock contemplating the yucca, my thoughts turned to nations and planets and solar systems and galaxies. There is no way, I concluded, for a nation or a galaxy to conduct its affairs from a basis of national or galactic pride and economic well-being, and still retain the mental and social attitudes we have come to value most in human societies.

What did you get out of this course? is a question very frequently asked, though not always in the same phrasing, when students are given the opportunity to evaluate an instructor. Is it possible to get too much out of a course? By *course*, we mean any learning experience. Is it possible to assimilate more than should be assimilated? Maybe; for I find myself every day now categorizing things in terms of love, joy, money, and pride. If I had to evaluate Steve Fretwell, I would say that he certainly was able, out on that hill above Keystone Lake, to teach a person to categorize things and activities in terms of love, joy, money, and pride.

We are sitting in the car now, my student and I, and instead of Ogallala or Keith County, we are in the parking lot of the student union at Kansas State University. It is early and we have driven to Manhattan for a meeting. The union coffee shop is not open, but we stroll through the lobby anyway. It is registration time; the lines are already forming for enrollment. It's the same all over the country, probably all over the world, wherever there's a university. Students line up to enroll. There is something different about this time and this place, however, and it's hard to put a finger on. Then it comes. The sign that could not be understood, the postings not really noticed at first, the IBM printouts posted all over a bunch of bulletin boards, standing all around where the lines were forming. Those bulletin boards are not a normal part of enrollment. Somebody has taken all the student teaching evaluations of every faculty member and posted them in public, in the student union, for everyone to read before they walk in to enroll. There is a sense of love and joy about this event. There are other places, where the evaluation of teaching is a matter of pride and money, where such computer printout results are totally confidential and no one knows what use is ever made of them, if any.

"Let's look up Fretwell's," I said.

"What do you bet they're pretty good?" says the student; he's been to Keith County. We look them up; posted right there on the bulletin board are Steve Fretwell's evaluations. They are excellent. We knew they would be. He has a love and joy for his work and demands love and joy of his students. Can't go too far wrong studying the dickcissel.

There are pride and money things that we are involved in, back in the capital city, for all things we are involved in are in fact centered in the capital city. Even the love and joy ones are administered back there, and the pride and money things are command performances. The forces of our subsociety, the academic profession, demand that we perform in the arena of pride and money, for there is some mistaken feeling in the capital city that those great institutions that now have much pride and much money have always worked for pride and money and out of a sense of pride and money. Our pride and money things involve the relationships between a cell and a protozoan that lives inside the cell. We are proud of the research we have done on these one-celled animals that must, I say again *must*, under certain conditions reside within another cell. We are proud to be a part of the elite group studying these small things, for these small things perform one function that is perfectly amazing. The intracellular environment that they require is inside the one-cell type, a macrophage, which is designed to digest foreign invaders! Our protozoa are efficient; they are able either to circumvent or to neutralize the digestive mechanism of their host cell. Our work has yielded money: we have a federal research grant and have had it for a number of years. The federal research grant does many things for a scientist. It allows him to travel, to pay for his scholarly publication costs, to buy needed chemicals and glassware, but most of all, as far as his *job* is concerned, it plunges him into an activity that will almost ensure he is an adequate teacher—his research. If a student's parents had the choice: learn to write poetry from a person who writes poetry or from a person who has learned how poetry is supposed to be written; I think the former teacher would be chosen every time. If a student's parents had the choice: learn to paint from a person who paints or from a person who has studied all history's "great" pictures, I think the former teacher would be chosen every time. It is easy for a student to determine whether a poetry teacher writes poetry; it is easy for a student to determine whether a painting instructor paints. It is not so easy for a student to determine whether a scientist does research. It is easy, however, for a student to determine whether a scientist's research is conducted out of pride and money or out of love and joy. It is *most* easy to determine whether a scientist's activity, conducted out of pride and money, has accidentally yielded some love and joy!

I am often asked, "Which do you prefer, teaching or research?" The questions are interesting to me, because they remind a person of two things. First, that the general public has little or no conception of what either teaching or research is, and this in itself is strange, because members of the general public are high consumers of both. Secondly, that in practice, most major universities are places where literally millions, millions, millions of dollars of taxpayers' funds are committed, where millions and millions more, the fruits of family labors, are spent by parents, where the responsibility for selection and "education" of those who will be our leading doctors, lawyers, politicians, and businessmen is placed, but where the people ultimately charged with the responsibility of all this commitment are evaluated on their ability to do something they were not hired to do. Excepting those who are actually hired to do research, for example, who are often agricultural experts or holders of privately endowed positions, the vast majority of faculty members at major universities, or minor ones for that matter, are hired to teach but evaluated on their ability to do research. This practice is the rough equivalent of hiring a person to work for the Department of Roads to count cars or fill in chuck-holes and then evaluating that person on his or her ability to sell shoes. If we were spending as much on the Department of Roads as we are spending on the university, and if citizens were paying as much annually in tolls as they pay for college educations, then you can bet Department of Roads employees would be evaluated on their ability to build and maintain roads and count cars. The point of my comments on the second of these phenomena in fact relates to the first phenomenon, the public understanding of the differences and individual characteristics of teaching and research. Research as it is visualized by the public is about as related to teaching as selling shoes is to counting cars. Research as it in fact exists is as related to teaching as cars are to counting cars.

Fretwell's contributions to the philosophy of teaching and research were very important in that they relate to the reasons for doing research, as an individual, and the part of biological research that is actually taught, by an individual, and thus to the differences between teaching and research. Pointed out again by a lecture that was so personal I thought it was terrible at the time and was embarrassed for Steve, out there on the hill above Keystone Lake, these differences between teaching and research were obvious as the structural and functional differences between an animal parasite and its animal host. While there may be vast structural and functional differences between a parasite and host, one must always realize that the parasite is dependent on the host. There are times when parasites kill hosts, but this is not the general rule. The continued existence of the parasite as a species is pretty dependent on the good health of the host. On the other hand the host may owe a small debt of gratitude to the parasite, since the latter may be one of the factors in control of the host species' population.

I mentioned earlier that if an activity should proceed out of love and joy but society demands that it be conducted according to the rules of pride and money, then something will be wrong. If research is the host and teaching is the parasite, then research that is conducted out of a sense of pride and money, but that *should be* conducted out of a sense of love and joy, will be an unhealthy host. Steve may have been feeling some pride and money pressures at the time. I don't know for sure; we didn't talk too much about his job or institution. But although he may have been feeling some pride and money pressures, he had been doing a great deal of love and joy thinking over the years. His point on the hill above Keystone was that there is probably no research that should be conducted according to the rules of pride and money. He was kind in leaving us with only an *impression* that there is no research whose existence should depend on or be maintained because of pride and money. I think, listening between the lines, that he was telling us loudly and clearly that over the long haul the love and joy research is all that ever filters through into the world's teaching activities, and in fact the part of that research that filters the furthest is not the results, but the approach. In its purest form, after all that filtration, the lasting approach is that of love and joy. What should be taught is love and joy, in fact, said Steve Fretwell; teach your students to love the total world of life and to get joy out of the revelations of any tiny part of it, and you have filtered and distilled the essence of knowledge and have passed some of it on.

I wash, almost every day, what seems like an awful lot of glassware. I cook up, for our one-celled animals, what seems like an awful lot of culture medium, which smells exactly like good soup. I spend what seems like an awful lot of time watching over students, the kids. I get up in the morning and go to work to wash dishes, cook, and take care of the kids. I get much love and joy out of this housework in my laboratory, especially washing the glassware. There is something about washing glassware that is very soothing, that occupies the hands but frees the mind. Most of my best research ideas have crept into my mind while washing test tubes. Flasks are not quite so productive as test tubes. The screw-cap culture tubes are especially productive. There are times when I simply have to let these tubes pile up in the sink, because washing them fills my head with too many exciting research ideas.

Back to basics is what glassware is, back to the basic part of being a biologist that drives a person there in the first place. My first recollection of a biological event was as a three- or four-year-old kid. I know very well that was the time, because I remember that back porch of that house so well, and particularly I remember my mother's rage and disappointment on that day. If you assigned a three- or four-year-old kid the task of catching a lizard, a green *Anolis*, on the back porch of his home in southern Louisiana, then that kid would probably not be able to catch the lizard. I have no idea how the lizard got there; I simply remember squeezing it until it died. My mother exploded with sorrow and disappointment. Three- or four-year-old kids can be very impressionable. A mother who explodes with sorrow and disappointment over the death of a lizard at the hands of her son conveys a pretty substantial message of love for living things and joy at their place in nature. It's been all downhill from there. Looking back, there have been decisions and decisions, but all of them have been to study biology, mainly animals. This is why I am here, to study biology and be a biology teacher. This is why I became a university professor in the first place, because the university is where a person studies animals and teaches about animals. Thus the glassware bit. The glassware bit is real; you wash the tubes, then put them in a rack, and before long there is a rack of clean tubes. These are partial environments for your animals; you are preparing environments for animals; your activities are directly related to *their* needs. The rack of tubes is reality. The committee meeting you excused yourself from just now is surreal. The committee meeting concerned pride and money, mainly pride.

A law was passed a few years ago in this state, declaring that the institution of higher learning in this state would be excellent. The state legislature passed the law. Along with the law came some money, and it was not long before a distinguished committee of "outsiders," including a representative from the very legislature that had passed the law to be excellent, was formed to determine if the institution had conformed with the law. This story now transcends a state or a university story, because it is a model situation.

Consider the legislation, spawned by the image of low intellectual ranking, whatever that might mean, within an "athletic conference," which itself has an image of low intellectual ranking. Such legislation to become excellent can only be passed because of affronted pride. There is the possibility that a law to become excellent was passed not because of affronted pride, but because someone convinced someone else that the latter's pride *should* be affronted. The law that was passed to alleviate the feelings of affronted pride was an appropriations bill. Pride and money. It was passed by a group that historically has a reputation, at least, of having no love for the very institution they legislated into excellence, and of certainly, very certainly, having no joy at the expenditure of money. Out of affronted pride, the group gave someone else's money to an institution they may not have loved and in the act of giving received no joy. The responsibility for converting this joyless financial act, performed out of questionable love because of affronted pride, into reality fell upon the shoulders of people like my friends. A flurry of activity ensued. Then a flurry of hyperactivity ensued. The joyless legislative act making it unlawful not to be excellent in turn spawned, as one might well guess, a set of activities conducted from a basis of money and pride rather than love and joy. We were given money and told to do something we would be proud of. Something went wrong.

Pride and money have historically characterized the arena of politics, so it should come as no surprise that a political body would act with money out of pride. Love and joy have historically characterized the arena of the mind, although there are many instances of love without the joy, but nevertheless love, in the creative and intellectual spheres. The mission of the soon-to-be-excellent institution was a mission of love and joy. No one in the history of that institution had ever said, "Memorize the reaction catalyzed by succinic dehydrogenase. Then be proud that you have memorized it, so that you may receive some money." Nor had they said, "Memorize the reaction catalyzed by succinic dehydrogenase. Then you will be so proud of yourself you will pay the institution some money." No, but down through the years some representatives of the institution had said, "There is a certain amount of joy to be derived from understanding the role of succinic dehydrogenase; you don't have to love this enzyme or this role, but you should at least be willing to admit that someone might love this sort of thing, because there are some sorts of things that you as an individual love." The institution was given three years to comply with the law. Both the institution and the legislature neglected to consider that the great institutions that already had much pride and money had generally operated out of love and joy for tens, sometimes hundreds, of years.

Then we went to Keith County, and that first day we collected some snails and saw the marsh wren. There were some of us in that crowd that first day who knew instantly where we were and where we had been. Those some had not a little difficulty leaving the marsh. One step into the marsh and reality returned. The flurry of activities intended to comply with the law had been everything but biology. The group had spent years learning how to learn about the world of life, then had spent years more doing everything except learning about the world of life. And finally, after doing everything for years except what they loved to do and got joy out of doing, except what they were trained to do, except playing their role in their time and their place, they were now asked to spend some money and generate some pride in order to comply with a law to become excellent. I sense it was the final straw. I sense that after ten or so years of trying to do for pride and money what they started to do out of love and joy, they were vulnerable to the marsh.

"Come home," said the marsh, "come back to where you belong." And the call was echoed from the South Platte River and Arthur Bay. They stood among the cliff swallows, and they tried very hard to imagine a committee meeting, and for the first time in years the committee meeting seemed *un*natural, a bad dream of tall old windows and cigar smoke.

"Try me," said Whitetail Creek. And the invitation was echoed by Ackley Valley Ranch. They stood at the headwaters of Whitetail and tried to imagine what it was they had been doing for years, and it all seemed *un*natural, a bad dream of graduate college forms and newspapers full of football. They became hardened quickly, and in the bars late in the afternoon they talked of fish and worms and snails and didn't really give in very much to forms and meetings and who-got-promoted and who-passed-comprehensives. Then they invited Steve Fretwell to come to Keith County for a week, and he came in his purple corduroy jeans and talked about all the things they had been sensing, feeling, but had been unable to express. He expressed it very quickly. A scientist searches for the truth about nature wherever that search may lead, but a scientist also searches out of love and joy, for pride and money will not long sustain a search for the truth. What had for years been a luxury, contact with the real world of nature as it lived in some place like Keith County, now became a necessity.

"Study nature, not books," said the man so long ago; so they began to study nature, not books, and they had been away from nature for so long that at first they fumbled, at first it was a little difficult, at first they had forgotten what it was like to study nature, not books. They began searching for nature, but they didn't search for nature quite the way a regular person would search for nature, because somewhere back in the files of their minds were the observations and feelings and approaches and attitudes that had served them so well the last time they had studied nature. So they took those things out of their files and found them slightly out-of-date but nevertheless valid, and they knew exactly where to look for the nature Steve Fretwell had told them to study with love and joy, and they found it at the man-nature interface in the marshes and streams of Keith County. Man and nature came together.

12

The Interface

MAN MEETS CARP
It was at the mouth of Sandy Creek, where the creek empties into Big Mac. It was morning, and the snail hunt was on. The instrument was glory sport equipment: a seventeen-foot *Boston Whaler* with an eighty-five-horsepower Evinrude, to hunt snails. The human hand and the human eye were also involved.

There were three of them, and they were running along the beach with gigs. One's gig had a string or cord attached, maybe a rope. They had a dog with them and they were yelling. Their average age was about eight years, and as the boat approached they ran into the water, yelling and waving their gigs, until something from one of them brought the other two. They converged on a spot near some dead trees and jabbed with their gigs. The carp was hit by one of them and finally beached. Immediately they began to gather wood for a fire. They were preparing to cook the fish. The kids had met nature in the form of a carp and had made the decision to eat it.

The water became too shallow for the *Whaler*, and we waded into the mouth of Sandy Creek. It was one of those places where the trees are all dead; acres of dead trees in the mouth of a Sandhills creek. Still there was no fire, and we wondered if there were any matches among them. When we left, one was still holding the carp under his arm like a package. He had not hung the carp on a dead tree branch by its gill operculum, as had dozens of fishermen before them. The carp decorated the dead trees like Christmas ornaments in varying states of decay. Dozens of fishermen had met nature in the form of a carp and had left nature hanging on a dead tree branch. On this trip, no one had met nature and decided to simply crush the carp, leaving them to sink with smashed air bladders. Nor had anyone simply stabbed a hunting knife through the bladders for the same purpose.

As everywhere, the dead carp littered the beaches near the mouth of Sandy Creek, but still, without the stabbing and crushing, all the interdigitations between man and carp had not been manifest; there was an interface, but it was not complete. The places where man and nature come together are so often marked by killing, but rarely the killing of man; and much of the killing also involves eating, again rarely of man. There is the killing of carp for eating. There is the killing of carp because carp are carp and compete with "game" fish, which also happen to be less bony. There is also the stabbing and crushing of carp.

One might think that the crushing of a carp's air bladder has some redeeming social value; at least the carp are supposed to sink and therefore not litter the beaches. Sunken carp may also be food for other things. Thus the crushing and stabbing of carp can be rationalized without much difficulty. That is, until one actually watches some people crushing and stabbing some carp. A person crushing a carp does not reveal much interesting behavior. A person crushing several carp reveals not only more interesting behavior but also a progression of behavior as he or she goes from fish to fish. A person crushing several carp is only slightly less interesting than a person crushing and stabbing several carp. One on one, person on carp, and there is a businesslike air to the crushing. Several on several, and we have a religious war to save the North Platte drainage from the carp. Next year, or maybe the year after, just as an experiment, I will single-handedly stop a war. The war will be abruptly stopped in the pitch of battle, and the individual soldiers will be asked to explain the purpose and value of their activities. Unfortunately, only the soldiers of one side will be able to offer their purposes and values. Some of these same soldiers, however, also carry assorted purposes and values into a war against the walleye.

The walleye war is normally conducted out of places like Otter Creek and the Lakeview Fishing Camp and involves thousands, perhaps millions, of dollars of sophisticated equipment. There is great dignity to the walleye war, and one senses that in this interface between man and nature, the fish provides much of the dignity. The walleye is excellent eating. One should realize that in the man-walleye encounter, physical contact between man and nature, that is, the fish, only enhances and sharpens the behavioral attributes of humans engaged in any phase of the war. Thinking about physically encountering a walleye produces some identifiable alteration in human behavior, and the alteration occurs generally in the direction *humans* interpret as dignity. Thinking about plus attempting to physically encounter the walleye adds measurably to the dignity of the activity, and furthermore, transforms the activity from "daydreaming" to "fishing," the latter, of course, being totally acceptable, almost required behavior on the North Platte. The former is considered slothful in most societies.

There is no book written, nor will there ever be any book written, that truly explores all of the relationships between man and fish. The interface between man and fish is so complicated and extensive as to defy comprehension. As a start, however, the average citizen should buy a boat. Simply as a mental exercise, if you do not already own a boat, add up your savings, your checking account balance, your convertible assets, then go out and *price* a boat. Better yet, price *several* boats. Start with a used twelve-foot aluminum job with a five-horsepower motor purchased through the want ads. Continue up the scale of price and sophistication until you reach the *Boston Whaler*, for that's about all that's *required* to adequately fish a lake twenty-two miles long and five miles wide. Be sure to add the eighty-five-horsepower Evinrude. And the trailer. And a car big enough to haul it.

Now, simply to add perspective and establish the educational background necessary to interpret your observations, soon to be made in the middle of Big Mac off Otter Creek (but generally applicable anywhere in this county), price several boats that are clearly beyond the level of sophistication *required* to fish for walleye, that is, those with some sort of cabin or awning, inboard-outboards, inboards, and so on. With your price list, go to the Otter Creek ramp on a Saturday morning and add up the dollars being floated on Big Mac for the sole purpose of fishing. It will boggle your mind. It will boggle your mind especially if you add to the interface between man and walleye the price of gasoline (car and boat), life preservers, licenses, paddles, beer, a camper-pickup or larger recreational vehicle to haul the boat, fishing rods, reels, line, depth finders, dip net, the salaries of game rangers, the cost of "wildlife management," the intellectual cost of "wildlife management," bait, lures, the dollars required to placate your wife or husband in order to get away with "fishing," the time you spend talking about unsuccessful fishing with your friends when you could be talking about *Stagnicola elodes* or cliff swallows (how much is your time worth to you on a per hour basis?), the time spent talking about successful fishing with your friends when you could be talking about *Fundulus kansae* and the South Platte River, radio advertising of the fishing camps along Big Mac, advertising of all kinds related to items necessary to fish for walleye, the salaries of professional fishermen that are maintained by the lure companies, the travel and equipment costs of professional fishermen. The list is probably longer. The point is pretty simple: man meets fish in a lot of ways. Man maintains the interface between man and fish and in so doing, also maintains, feeds, and clothes many men.

Europeans and Asians also maintain an interface with the carp, but it is an interface characterized by deepest respect on both the hatchery dike and the dinner table. In Europe and Asia the man-carp interface feeds and clothes many men.

MAN MANAGES DUCK

Man manages duck in a place called the Crescent Lake National Wildlife Refuge out in Garden County, not far from Ogallala. Crescent Lake, as the refuge is affectionately called by those who've been there, is a refuge not only for waterfowl, but for humans. A biology student, contemplating a career selling life insurance, wonders what "kinds of jobs" might be available for a graduated biology student and is often told, "Fish and Wildlife Service refuge jobs." Armed with this information, the biology student then contemplates life with a wife in places like Crescent Lake and the Aransas National Wildlife Refuge out of Refugio, Texas.

One gets to Crescent Lake by going north from Oshkosh many miles along a dirt road. There are a number of lakes in and around the Crescent Lake refuge, which have names like Swan Lake, Goose Lake, Hackberry Lake; and they are places where the aquifer is clipped by the dips in the sandhills. There is a fire tower at Crescent Lake, and if one is on official business, the rangers may unlock the fire tower and allow a climb. From the top it is desolation as far as the eye can see, a most serene and beautiful desolation. There are very few trees, no houses except at the refuge headquarters, a few trails. Clouds cast gigantic shadows on the sandhills. The lakes are smooth and still. Down below, halfway up the tower, there is a kingbird's nest in the corner of the steel framework that holds the observation platform. From here you can see the eggs; the kingbird left when you started up. This day, also, there were barn owls nesting in the fire tower platform, so we were not allowed up into the observation box itself. A third of the way up, the owl had left. Above our heads it slipped through a broken window and went arrow straight and silent across the land out of sight. It is an uncomfortable and disturbing feeling, up on the fire tower.

Even with several people on the fire tower, the sense is that of total aloneness, of intrusion, of voyeurism almost, as one looks over the land and water that are so obviously doing well without him.

Once, down on the road, we stopped and looked over Crescent Lake with the binoculars. The land's heat and mirage distorted the mirror surface and line of marsh over the flattened distance. There were ducks on the lake, hundreds of them, in no particular pattern, and their images wrinkled and waved and simmered across the mile of hayed prairie. The event happened so quickly

Western Grebe

"It was the courtship dance of the Western Grebes. Come and gone like a falling star. One can never get enough of a falling star."

it didn't really register until that moment later when I tried to decide whether it was real, a part of the mirage, or simply my eyes in the heat. Against the distant line of the marsh had briefly appeared a puff of white, a churning. There were no boats on the lake. It was the courtship dance of the western grebes. Come and gone like a falling star. One can never get enough of a falling star.

Refuges are great places, and the idea of a refuge where ducks may breed unmolested, protected by federal law and federal bureaucracy, isolated from the predations of children and human poachers, is satisfying. However, there is not a single duck refuged and managed and bred, with the aid of human kindness, that is not intended to be killed with a shotgun. One immediately wonders: If all waterfowl hunting were permanently stopped, would Crescent Lake's budget also be permanently stopped and would Ducks Unlimited continue more than a few months? The individual cost of duck hunting is virtually prohibitive, so that the per-pound-of-meat cost may run into the hundreds of dollars. No one spends even fifteen dollars a pound for table meat, so duck hunting is indeed a sport rather than a means of subsistence. One also wonders how long Crescent Lake and Ducks Unlimited would last if no duck hunter was ever able to kill a duck or if federal law allowed only blank ammunition in the shotguns.

Blank ammunition would be an interesting experiment. One would still get all the thrill of buying decoys, carrying decoys, getting stuck in the mud in early morning hours, watching the sun rise, building a blind, testing skill at setting decoys, freezing your ass off, testing skill at calling and identification (identification skill is an absolute requirement nowadays for staying out of jail), talking about duck hunting, seeing all the fringe benefits of duck hunting (like coots and mergansers and western grebes), carrying a gun, loading a gun, and firing a gun. Is the actual kill essential? Is it in fact necessary for the animal to be blasted? Might there be a role for a Labrador retriever other than going into the water after a *dead* duck? The questions of course will never be answered because there will never be a federal blank ammunition law. The refuge will still be a place where every human effort is made to encourage the successful breeding of ducks and geese that can be killed, actually killed, for human entertainment. It would be politically unwise for any elected public official to come out against hunting. Humans will hunt, if not for meat, then for entertainment. Humans will not generally hunt for very long without killing something, if not for meat, then for entertainment. It is a shame, however, that a totally innocent animal like *Stagnicola* is caught up in this federal organization-taxpayer entertainment network. In its efforts to breed a population of waterfowl large enough to maintain nationwide killing for entertainment, the Fish and Wildlife Service has brought a plague on snails.

When ducks descend upon the Crescent Lake Wildlife Refuge, they bring parasitic worms with them, in their intestines, from more northern climes in the fall and from South America, Mexico, and the Gulf Coast in the spring. Many, many ducks are infected with these worms, and there are many kinds of these worms, some of which cause severe illness in the birds, and all of which participate in the ducks' feeding ritual by taking materials directly from the birds' intestinal contents. The duck population has been dropping worm eggs in every puddle and river along the way, and will continue to drop worm eggs until the duck reaches its final migration destination. Or until the worm dies, which does sometimes happen, since northern worms are not always able to survive on southern duck food, and vice versa. But today the hoards of waterfowl are dropping worm eggs all over Crescent Lake. A single duck may drop millions of eggs in its feces over the course of a nesting season.

The life of a worm is difficult for a student to learn. The life of an intestinal worm, parasitic in a duck, departs from reality enough so that a student has some difficulty believing what is known. A worm egg, a trematode egg, either hatches in Crescent Lake or is eaten by a snail, depending upon the kind of worm; but in either case, whether the egg hatches in the water or inside the snail, the result is a small animalcule. This creature burrows into the snail's liver and begins to multiply, producing in the process several different stages of maturity. If one crushes an infected snail, one may find hundreds of these worms, some exceedingly active, all pulsating with the curious but frightening movements of parasitic worms. Over the course of its life, a single snail may produce hundreds, thousands, of these larval worms, ready to infect a duck. One only has to crush the snail to find the worms; they spill out into the dish with abandon, writhing. It is not too unusual to find more than one kind in a single snail. Nor is it unusual on the Crescent Lake refuge to be able to pick up ten pounds of snails in a very short time. Thirty minutes in Goose Lake will net maybe five hundred, each as large as the terminal joint on your little finger. In Goose Lake every snail is infected with parasitic worms and some are infected with more than one kind. The worm population being maintained by the Fish and Wildlife Service is astronomical, well beyond anyone's estimation. There is little doubt that baby ducks, off the nest quickly after hatching, get infected with virtually their first meal. There is little doubt that in concentrated breeding areas all over the nation, in places where waterfowl are concentrated and bred and exported for entertainment killing, the parasites are also concentrated and bred and exported. Your immediate reaction concerns the welfare of the ducks: do these parasitic infections harm the ducks so much that we are not getting as many adults as we could for the entertainment of the hunters? Is our tax money going to waste because the infections are not controlled? Probably the answer to both questions is a qualified "yes." On the other hand, no taxpayer ever asks if the *Stagnicola elodes* population is being infected, its collective liver being eaten out, being forced to support a very serious disease burden, because of this arrangement for producing ducks for hunters? The answer to this last question, of course, is an unqualified "you bet."

KINGBIRDS DEDICATE A SCULPTURE

There is a phenomenon, in the state, known as the "controversial I-80 sculptures." Now that the sculptures are up, of course, there is no controversy, and indeed the only controversy that ever really arose was generated by a very few individuals who didn't think abstract sculpture was an appropriate way to celebrate the bicentennial. The newspapers bent over backwards to let these individuals have their say, and in doing so created a "controversy" that was obviously news in itself. The "controversy" was fanned by the media until it forced the state legislature into the issue with public hearings at the sculpture sites. Public hearings by a legislative committee on a "controversial" issue with testimony from every important person in that state, including the director of the Department of Roads and the president of the university system, as well as every local person of stature in the outstate community, is indeed big news. In the fields of biology and engineering, this scenario would be a positive feed-forward set of events.

The piece at Ogallala is called *Up/Over*, by Linda Howard. It is one of several in a four-hundred-mile-long sculpture garden, and it was dedicated on July 4, 1976, at the rest-stop on the north side of I-80 between Ogallala and Roscoe. The eastern kingbirds were in the small trees planted only a few years ago, around the rest-stop buildings. It was a recently fledged brood. I wondered at the time if anyone else noticed the birds, or noticed that evidently only part of the clutch had been successfully fledged. There

are few things more gentle than a recently fledged eastern kingbird, just as there are few things less gentle than an adult western
kingbird. The two fledglings flew very gently up/over the crowd.

The artist, Linda Howard, stood in front of the rest-stop
building with the photographer, Dan Ladely. Linda had long
blonde hair and was wearing a knee-length skirt. Dan looked
like a cowboy out of a Gene Autry movie, with ample but not too
ample girth. Two state patrol officers, resplendent, stood in front
of their cars, also resplendent, and smoked. One had a resplendent
boot placed carefully on his front bumper, and when he moved
there was a tiny mark on the bumper. The lieutenant governor,
Gerald Whelan, arrived by helicopter, was met by the rest of the
dignitaries, and gave his speech in front of *Up/Over*. Dan was
standing in back of the piece, taking pictures. The sculpture
itself is an arc of I-beams, and there is no way to look at it, no
angle, from which it does not fit exactly into, complementing and
enhancing the dignity of, the backdrop sandhills. *Up/Over* is
abstract I-beam sculpture, and of course the artist was not about
to explain to the crowd what the piece was supposed to mean, if
anything. It's a flock of pelicans, meeting a Sandhills lake or prairie
marsh; it's pioneers on Windlass Hill out at Ash Hollow; it's a
long-billed curlew; it's the historical adjustments that have led
to the present standoff between man and nature in Keith County;
or it's a kingbird dealing with the dedication-day crowd.

Art Thompson was a former program director at the University's
student union, and he moved his office into the Sheldon Art Gallery when he took his new job as director of the I-80 sculpture
project. The sculptures are Art's babies, and there are more than
the one at Ogallala. *Erma's Desire* at Grand Island may be one
of the most dramatic pieces of public sculpture ever created, and
much of the drama comes from the setting. There may not be
another piece of art in the entire world that works with nature
the way *Erma's Desire* does, that can tell the story of a whole
section of the world as one drives past, going east, at fifty-five
miles per hour. *Erma's Desire* was the most controversial work
at the time. The controversy now probably centers around whether
Erma may in fact be the best piece of sculpture in the world.

There is no way to accurately describe Art Thompson's role in the creation of a sculpture garden. On second thought, there is a way. A kingbird fledgling is alert, upright, quick but so gentle with the eyes, sometimes ragged on the takeoff and landing, but not very ragged, and always pressing the air, pushing it with quick but soft small wings, chattering slightly, a soft complaint at times, a touch of affront at others, and a sure-fire crowd-pleaser, every time no fail. That's also Art Thompson. That day the king-birds had landed in the tree above where his kids were playing.

MAN ASSAULTS ARTHUR BAY

Arthur Bay is located on a lake that is twenty-two miles long and five miles wide, with crystal beaches on both sides and spectacular isolated scenic spots. Arthur Bay is not one of these spots. The Arthur Bay road is a couple of miles west of Sportsman's Complex, and if one turns onto the Arthur Bay road late at night the kangaroo rats begin popping out of the grass as soon as the car lights leave the asphalt. There is a boat ramp and an outhouse at Arthur Bay, but one can visit Arthur Bay every day for a year, see something new every day, and never see either the boat ramp or the outhouse. Turning left off the Arthur Bay road, one very soon reaches an exceedingly secluded and mysterious place where the water is shallow and warm, and willows hide the bank, and Arthur Bay is thrown into many confusing folds, little bays, each with its own set of willows. Beneath the willows is the constant smell and sight of dead carp and flies. There are aluminum fishing boats, sometimes painted red, with small five-horsepower motors, wedged furtively up into the willows, no fishermen in sight. Sometimes there are campers, recreational vehicles, also wedged up into the willows. The trail is steep, very sandy in places, very unlevel, and sinuous. It is one of the better places to get stuck far from help.

If one continues along the main road into Arthur Bay, one arrives quickly at the boat ramp and outhouse, and the road is still wide, graveled and graded, and cut into the sandhills at places where kangaroo rats have reopened their burrows. The ramp is one of the lake's better ones, at least for two-wheel-drive vehicles, and the outhouse is functional and relatively clean.

One can, on one of society's calm days, or early in the morning, launch a sailboat from Arthur Bay without much trouble and with a little breeze negotiate the narrow channel and get into the main part of the lake itself. To clear the point out of Arthur Bay and take a close reach toward the south shore is to leave the world behind. To any sailor at any time, the first snap of the dacron as it catches that first air is the sound that draws the line between land and water. On land there is the city, the boss or immediate supervisor, and the car. On water, there is the boat and the wind. And the companion. On this day the companion was a girl, Crazy Diane they sometimes called her, and she had brought her breakfast, a six-pack of tomato juice.

There were guests in town, some students' wives, and we had been drinking the night before at Bill and Rhonda's place. Some folks like my wife Karen and me had simply sipped on a couple of beers while others dug into sizzling steaks. Bill and Rhonda's place is called the Sip 'n Sizzle. It had been a rowdy Friday night in town, however—the Friday of a long weekend. Our daughter had requested permission to also go drinking and carousing, and permission had been denied. Thirteen-year-old girls can't go carousing without their parents in Ogallala, so her parents had decided to go carousing without her. The students without wives were sometimes rowdy, particularly in and around the pool table, and particularly around the pool table if a local hustler was trying them. There are a number of local hustlers in Ogallala. We had returned late to the darkened camp. All was intact. A last-minute decision to placate the baby-sitter had been made: we would take her an order of Rhonda's special onion rings. Rhonda's onion rings cannot be described, especially as they taste right then, hot, with an ice-cold just-opened can of Bud. Evidently they don't have the same appeal to a thirteen-year-old baby sitter who has been denied the permission to go carousing and who has since gone to sleep. I ate the onion rings for breakfast.

Saturday, July 3, 1976: bright and perfect. Camp was deserted and asleep. Fidgeting, I started recruiting crew at the top. In order: Wife, thirteen-year-old-baby-sitter-daughter-who-usually-puts-pressure-on-to-go-sailing, other children (the baby-sitter was working in the kitchen until late that afternoon), male students, other scientists, the director (hates boats, loves baseball). Crazy Diane was the one person in camp who was ready to sail. She was sound asleep and I gave her five minutes. Six minutes later we were on the road. I brought the onion rings. We stopped at Kingsley Lodge for a carry-out breakfast: she bought the six-pack of tomato juice, I bought a six-pack of Pabst. The mast went up easily, and we negotiated the point at Arthur Bay as the campers and Colorado boats began to rumble in along the gravel road. A few sleepy people with sandy feet watched from their tents and the doors of their trailers as we tacked through the channel.

You can hear quite a bit from a sailboat. Later from far across the lake we could hear the big inboards. The sound of an inboard is a particularly vicious sound coming across the lake. It is an unyielding sound, and it brings to mind a thousand pounds of arrow hurtling across the water with the driver looking backwards at the fallen water-skier. In the middle of Big Mac it was sun and wind and dacron and tomato juice and Pabst and onion rings. It was only later, meeting Karen and the kids, finally awake on a lazy Saturday, back at Arthur Bay, that the meaning of the inboards struck home. That sound that had drifted out on Big Mac, that deep gurgle spitting into the machinegun staccato and accelerating into the solid roar, was the sound of Arthur Bay. My sailboat will not work in three inches of water. I am continually amazed that thousands of dollars worth of inboard will.

I could tell from a hundred yards out she was terrified. The way she walked, the way the kids ran in front of her, the way she glanced periodically in all directions, up and down the beach, out over the water—all said she's terrified. The inboard had passed within only a few yards of our daughter and a friend, swimming a few feet from the water line in the tap-clear water off the Arthur Bay point. It was going a hundred miles an hour in her mind, and the driver was looking backwards, an empty ski rope skipping behind, until a hard crank on the wheel spun the boat on its stern and looped the rope. We ran up onto the beach and I picked up the kids. Crazy Diane went back to camp. The wind was marginal but the lake was churning. A current of ragged waves was coming out of Arthur Bay, out the channel we had negotiated earlier.

Much later that afternoon, finished sailing, we dropped the kids on the point. The son of our friends was along, fifteen years old and a master angler. There was no doubt in my mind we could dismantle the sailboat and get back to camp. It had been obvious from late in the afternoon that we could not sail back into Arthur Bay, especially to the boat ramp near where my car and trailer were parked. The current was strong now, with waves that rattled the mast and the wind was light. The current was that of boats churning the waters of Arthur Bay, so that the waves spurted out the channel as a current. Alan and I looked for a long time at Arthur Bay from the point. We decided to walk the Lido; it pulls easily in knee-deep water with the center board and rudder up. We would walk it along the shore to the ramp, unstep the mast, and make it back for dinner. The plan almost worked.

There were the dogs, hundreds of dogs, all shapes and sizes, right out of a piece of children's literature or the encyclopedia. They were tied beneath the Winnebagos, they were tied to the volley ball nets, they ran loose, they barked at the dogs tied underneath the next Winnebago and to the next volleyball net, they got their chains tangled in the lawn chairs. Some had collars, some had tags. There were some small packs of dogs. They barked at us steadily. Some ran up to us in the water. There were big German shepherds, loping at the water's edge, and one always wonders who and what a German shepherd will become protective about. Their masters were almost always jogging, sweating along the beach a hundred yards behind, a terrycloth headband and an Iowa State University Athletic Department T-shirt above a pair of clashing red shorts. There were the Dobermans, and they often flashed out at the shepherds. It would not have been out of character for some large dog to bite in half some small child that day, but it did not happen. It might not have been out of character for some small child to bite a large dog in half that day, but either that did not happen or we did not see it. Neither would it have been out of character for either a dog or a small child to be cut in half by an outboard motor, but mercifully it did not happen that day. The only things more numerous than the dogs were the kids. We saw no cats. I privately wonder how many wild dogs, wolves, and coyotes came out of the hills on July third to get a couple of days of cheap food and beer.

There were at least fifty large boats in Arthur Bay. Their average speed was about thirty miles per hour and their average length of straight course was fifty yards. It would have helped if they had all been going the same direction, but they went in several concentric as well as overlapping circles, some clockwise, some counterclockwise, and all with skiers. Well, not all: there was the one with the five-year-old kid on an inner tube, being pulled at thirty miles per hour among the other boats pulling the skiers. Frequently they ran up onto the beach. Frequently they ran up onto the beach on purpose. Frequently the skiers fell. There was much yelling and cursing out on the water. Most of the drivers were looking behind them, at the skiers. It is interesting to watch fifty boats all going thirty miles per hour in a variety of circles in Arthur Bay all pulling skiers or kids or dogs on inner tubes, some falling off, and fifty drivers, all looking backwards at the fallen skiers. The amount of human skill displayed is amazing. What is particularly amazing is that all this can be done with a can of Coors held to the mouth. The kid on the inner tube was not drinking Coors. The gurgle and spit of the inboards cut through the sound of the outboards. The late afternoon sun began to cast some shadows of the bluffs that seal off the west end of Arthur Bay.

On shore the Frisbees and volleyballs flew. Alan and I looked down the beach toward the ramp. At any one time one could see a hundred people with beer cans to their mouths. We pulled the Lido through the water beyond where the campers had their beer in the lake. They were family to family. Their charcoal fires were smoking. Sometimes their volleyball nets divided the beach. They sat in lawn chairs, they ran, they chased dogs, they chased kids, and they chased volleyballs and Frisbees. They were having a great time, and they had no idea a hail storm would come close to canceling the fireworks display on Sunday.

Near the ramp we prepared the Lido for the trailer. Alan's father studies water and fish and is very very familiar with motor boats of all sizes and shapes. Alan is obviously far more practical and learned in the matter of motorized aquatic transportation than is a sailor. I had wondered aloud why no one skied out on the big lake itself, why of all places to go to ski and motor a person would choose Arthur Bay. He had looked at me with the disbelief that only an adolescent can muster for an adult, the disbelief that the adult's train of thought has not led to the obvious answer. "Takes less gas," he said.

Back behind us the four-wheel-drive vehicles were trying their skill at the sand bluffs that border Arthur Bay. Their whines blended nicely with the cut of the inboards and the more familiar sounds of the outboards. The sand bluffs are about twenty feet high and nearly vertical at the top, a challenge for the vehicles and drivers. It usually takes several runs to get over the top, and the four-wheel-drive vehicles either back down the bluffs at about forty miles per hour or back down partway then make a sharp circle in the wet sand. The kids of all ages on the beach, as well as the volleyball nets, add to the challenge, for the driver evidently must miss these in order to continue the assault on the bluff. A smart kid does not play on the beach between tire tracks. The traffic on the Arthur Bay beach on July third is slightly more intense and travels slightly more rapidly than the traffic on the main street through downtown Ogallala, and no smart kid would play on Highway 30 through town. It is not easy to see a child playing between the tracks on the Arthur Bay beach at twilight, especially if one is driving a large four-wheel-drive vehicle and drinking at the same time and watching for the idiots who drive their boats clear up on the sand. It's also hard to tell the difference between a big dog and a little kid as they flash in your rearview mirror, as you're backing down the bluff at full throttle. Most of the four-wheel-drive vehicles out on Arthur Bay are pretty enclosed.

It was dark now, and we'd missed dinner, and the four-wheel-drive vehicles had their lights on, at least some of them, and we'd finally gotten the Lido on the trailer and up the hill. The four-wheel-drives had driven continuously across the ramp on their way to the bluffs, and loading the boat had been something like trying to load it across the traffic on Highway 30 through town. Even that would not have been so bad had it not been for the people cutting in front of us with their boats or trailers. The cross-traffic and cut-ins had a pattern, however, so we eventually managed to work around and through those. The guy who launched his wife and inboard with the drain plug out was almost the end of it all. His job of backing down the ramp from way up the hill through the cross-traffic was as masterful a bit of high-performance driving I've seen. It's one thing to go full throttle in reverse down a hill in a four-wheel-drive vehicle. It's quite another to do the same full throttle reverse down a boat ramp with the empty trailer aimed straight, dodging cross traffic and people standing near the ramp with their backs turned toward you. I was very very glad to be on the Arthur Bay road. I would have been glad at that moment to be back on the Arthur Bay road even had I known at the time that news of the accident was waiting back at camp.

Jena, the ten-year-old, is as true a dog lover as exists, and the fact that the accident had happened right in front of Jena was about all her mother needed to finalize her impressions of humanity's assault on Arthur Bay. It was an Irish setter and it was hit by a four-wheel-drive vehicle while both were bounding down the beach. There were sand castles where the dog was hit. It's also impossible to know how many *Bufo woodhousei* were ground into the sand beneath the tire tracks on Arthur Bay beach.

Arthur Bay assaults man

Bufo woodhousei is also known as the Rocky Mountain Toad. There are people in the world who love toads, although there are also some toads that are the most grotesque animals on the face of the earth. Furthermore, some toads are not only grotesque but act grotesque, so that only a true toad lover could love those species. Correction: only a true toad lover could love those species never having met *B. woodhousei*. *Bufo woodhousei* makes a toad lover out of anyone who steps out of the vehicles at Arthur Bay. One year little Doug, the redheaded two-year-old toad lover, spent five weeks with a *Bufo woodhousei* held tightly in a dirty little hand. The next year he did the same thing.

If one turns left off the Arthur Bay road into that very secluded and mysterious place where the water is shallow and warm and the willows hide the bank, one finds hundreds, maybe thousands, of *B. woodhousei*. They're not superfast and thus are caught easily. Furthermore, they tend to act very surprised that a human has come to Arthur Bay and don't make much of an effort not to get caught. This is blasphemy, of course, to write about an animal like a toad in romantic and unscientific words, to give the wart factory a personality, a set of motives, a set of values and approaches to life that only few humans aspire to. The animals may be incredibly stupid, they may be locked in a most vicious struggle with one another for survival on the Arthur Bay beachhead, they may hate humans, they may all be at this moment wishing they were not toads so they could be in town drinking beer with the college kids. Who knows what a toad thinks and feels? On the other hand, who of those that have been to Arthur Bay can come away not feeling that they have encountered a toad with a personality, a set of values, motives and approaches to life that few humans aspire to?

Even a newly metamorphosed *B. woodhousei* is a grandfather, but the older ones are especially grandfathers. Everything any grandfather has ever been to a kid is what *B. woodhousei* is to a human. Mr. *woodhousei* treats one with kindness, respect, and gentleness. He tolerates some abuse with an understanding look. He puts up with the catch-'em game because he knows the kids will soon tire and will stop pestering him. He also knows that humans' initial tendency to catch, kill, and abuse a wild thing will soon be assaulted and beat down, since no one, absolutely no one, can abuse a *B. woodhousei* forever. Arthur Bay puts its best up front on the table: one step out of the vehicle and the human is catching toads. It doesn't last long. It might last the first day out, but not through the second. One cannot get through an hour of catching toads without asking why he is catching so many, of what possible use these gentle animals could be, how many actually need be killed in order to determine whether there are worms in the intestine or parasitic protozoa in the blood. It's not long before the same humans are asking how many fish actually need be seined out of Arthur Bay, even how many metallic flies need be swept off a dead carp. Not many humans go back to Arthur Bay, having met *B. woodhousei*, and take more metallic flies than they actually need. Not many humans, having been to Arthur Bay and having met *B. woodhousei*, go anywhere and take from nature more than they actually need to teach another human or to discover those essential relationships between parts of nature.

Arthur Bay is the home of *B. woodhousei* in the strongest of senses. The black tadpoles swarm in the large rainpool off to the left of the Arthur Bay road, and the adults are everywhere. One knows they eat insects, and one knows certainly that there are plenty of insects at Arthur Bay: metallic flies, tiger beetles. But one also senses that *B. woodhousei* doesn't eat more than it needs. One also senses that *B. woodhousei* really doesn't object to a stranger's coming into its home. *B. woodhousei* will share Arthur Bay with you, even if you grind him into the sand with your four-wheel-drive vehicle. *B. woodhousei* will share Arthur Bay with you even if you kill him to find out what's inside his intestine. *B. woodhousei* will share Arthur Bay with you even if you catch his flies, put him in a cage to starve and dry to hardness, or step on him accidentally. He will be there again next year if you don't learn your lessons well enough this year. He knows that all you have to do is stop and consider him as what he is, a toad, and there is no way you will be able to stop the process of absorbing his values and approaches. He knows he can teach you, with his own life if necessary, not to take more than you need from Arthur Bay. He knows that you will discover the beauty in his style, which surpasses many times the beauty in his face. He knows those lessons you learn by abusing him at Arthur Bay will stick with you when you go to Martin Bay, or Clear Creek, or

Whitetail Creek, or Keystone Marsh. He may also be hoping that one of those attractive coeds in cutoffs and halter top, thinking he's a frog, will give him a kiss. There is no way one of these girls will give him a kiss, however, since they are biologists and know full well he eats flies that live on dead carp.

There are at least several kinds of flies on dead carp, and of all the kinds there are at least two that are shiny and metallic. The two species belong to the genera *Paralucilia* and *Protocalliphora* and are metallic green and blue respectively. They give of themselves as readily as does *B. woodhousei* but are evidently not blessed with quite as many behavioral options, for they are warm weather friends, *Protocalliphora* more than *Paralucilia*. Their larvae can reduce a dead carp to nothing in a very short time, or at least the larvae of one of them, maybe, because we really don't know if that seething mass of maggots within the carp came from the eggs of one of these flies.

The experience of actually picking up a dead carp, on a reasonably still day, and poking through the maggots is one of the most olfactory experiences a person can have. There are few— very very few—smells like it. The smell is repulsive and sickening even to a parasitologist who will regularly pick up a dead raccoon, hours old, off the highway on a hot day and slit open its intestine to find the worms. It is even more of an experience to collect a maggot-sand-carp jelly sample and take it back to the laboratory. There is little doubt, having brought the sample into the laboratory, that this set of fly larvae, bacteria, and sand belongs left alone in nature.

The flies are not subtle; they are quick and solid of body. The adults can live for days and days unfed in a plastic bottle in the refrigerator. Well, some of them can live for days and days. Normally some will die and the remainder will lay eggs on the dead brethren, and the plastic jar will then be full of not only living adults but small maggots. The flies are members of the same family, Calliphoridae, and of course not only will they remind you that they belong left alone in nature, but they will shame you with their singleness of purpose and the businesslike pursuit of that purpose (assuming purpose is a virtue). They assume their positions on the dead carp; not just any dead carp, mind you, but one that is just the right consistency, the right degree of dryness, the right distance from the waterline along the sand beach. And is dead on the right day, for the composition of the fly population on dead carp varies with the weather. The various species obviously have their preferences, but the green ones are more hardy.

The day was terrible for flies. The wind was blowing maybe fifty miles per hour and had been for three or four days, so that the local entertainment had become driving up to the dam to see if it was ready to break under the pounding of massive waves. It was cold, for summer, and only *Paralucilia* was found on the carp. This was particularly distressing. There are certain times a teacher would like to be successful and, above all, right. Some of these certain times are when the subject is a group of protozoa upon which that teacher is supposed to be an expert, a group of protozoa that live inside the intestines of calliphorid flies. The touchiest of these some certain times occurs when in addition a cell biologist teacher has been extolling the significance of an ecological question: why are *Protocalliphora* so heavily infected with these protozoa while neither *Paralucilia* nor the unidentified maggots within the flesh, all from the same dead carp, are not? These last ones are the times when it's satisfyingly macho to be able to call one's shot. These are also the times that Arthur Bay can jerk one back to his senses.

There were plenty of flies, all green. Back in the lab one single protozoan was found out of two or three hundred flies. Credibility suffered. Some ecology problem. It mattered little that the flies were all *Paralucilia*; the significance of the genus name is lost unless there is some obvious functional difference between a member of that genus and a member of another genus. We were looking for protozoa and there were none. So much for that. We tried it and it didn't work. Maybe next year will be a good year for flies on dead carp. There is little doubt that next year Arthur Bay's flies will be approached with considerably more skill and

[189]

forethought and considerably less macho. There is reason to expect that even Martin Bay, or Clear Creek, or Whitetail Creek, or Keystone Marsh will also be approached with considerably more skill and forethought and considerably less macho.

The kangaroo rat, *Dipodomys ordi*, is one of the more beautiful animals on the face of the earth, beautiful in form and color and texture as well as in behavior, for individuals are notoriously gentle, serene in captivity, and not known for biting. *Dipodomys ordi* is also a potential seething bundle of nerves, for several together in a wire cage soon rip one another apart. It is significant that one never sees several together in a wire cage in nature, nor does one see several togther in a terrarium or zoo habitat, unless there is adequate room and facilities for the establishment of territories and ranges. *Dipodomys ordi* is also nocturnal, which a human is not, so to meet *D. ordi* the human must either become nocturnal and go to Arthur Bay at night or else he must set traps. In Keith County one chooses to go to Arthur Bay at night.

One can go to Arthur Bay at night in either a state-owned vehicle or a private car; each has its advantages. If one has gone to Arthur Bay in a private vehicle and *D. ordi* is not out, then one can go directly from Arthur Bay to a local tavern. The state-owned vehicle, however, is much better suited for meeting *D. ordi*. It has large doors and large running boards upon which people can ride, ready to leap with flashlight and net after *D. ordi*, and lots of room for individual cages. One doesn't drive a state vehicle to a local tavern, however, so if *D. ordi* is not out, one stands around on the Arthur Bay road for a while kicking rocks and then goes home. Then one drives a private car to the local tavern.

Dipodomys ordi exhibits the structural adaptations of all rodents that live on sand. Bipedal locomotion, long brush-tipped tail for balance, large ear capsules. There was a time early in the days of space-exploration thinking when *D. ordi*'s large ear capsules made it an ideal experimental animal for studying the mechanisms of balance. Perhaps our astronauts balance better because of that research on *D. ordi* and its inner ear. With or without the research, *D. ordi* balances beautifully, sitting or full steam ahead. Full steam ahead is pretty fast, and one must conclude that *D. ordi* does not receive its full load of chiggers and lice while going full steam ahead through the yucca and prickly pear of Arthur Bay.

A legitimate bloodsucking louse is a thing of rare beauty. Animals all over the world are covered with them, including many human beings. Thus it is always a surprise when people are surprised to find a legitimate bloodsucking louse on some creature. The surprise is somewhat understandable in the case of *D. ordi* however, since the rodent is so beautiful the louse's own rare beauty is quite overshadowed. One does not think a louse should be on a kangaroo rat. On the other hand, it is difficult to see a louse without very soon thereafter summoning up an immense amount of respect for the insect. Any louse will do, as long as it is a legitimate bloodsucking louse. It is a curious fact that louse epizootiology is not taught in the public schools. But then neither is mosquito biology. There is a great deal of literature, such as that written by Hans Zinnser, that tells us convincingly that the earth's political history should not be written in terms of military decisions, but rather in terms of malaria and typhus epidemics, neither of which would exist without anopheline mosquitoes and anopluran lice respectively.

Dipodomys ordi lends itself well to study of the louse-rodent interface, for the louse glues its eggs to the fur of the rat. The chigger, on the other hand, takes its chances out in the brush, awaits *D. ordi*, and grabs at the first opportunity. *Dipodomys ordi* also lends itself well to the study of the chigger-rodent interface. *Dipodomys ordi* lends itself especially well to the study of obligate associations between unrelated organisms, particularly since it participates in more than one such association and the two associations are maintained by two very different encounter mechanisms. Only one must catch *D. ordi* first, and in some reasonable numbers.

It is night at Arthur Bay, and there is no moon and a slight breeze. The waves on Lake McConaughy are breaking on the north shore a few hundred yards away, and camper lights are scattered randomly in the direction of the wave sounds. Periodically there is a door slammed, an outburst from some card game being played in a Colorado recreational vehicle by people drinking Coors out of Julesburg. Far off on some road auto lights rise and fall, dipping below the dunes and reappearing at some unexplained angle. One wonders if the dipping lights will eventually meander through the dunes to one's own road, and the thought does make a difference. The lights could be those of a midnight ranger, or worse yet, a taxpayer. It would be bad enough to end up in jail simply for trying to study the *D. ordi*–louse interface. It would be worse yet to be held responsible for a fatal heart attack of some taxpayer rounding the dune in a Winnebago only to find himself in the middle of the Keystone cops studying the *D. ordi*–louse interface.

The cops are experienced now, and one stays in the state van, foot on brake, while the others study the *D. ordi*–louse interface. The dust of the Arthur Bay road is still in the air, but the van lights stab far enough to pick up the struggle and chase. There are five of them, and they all have insect nets and powerful flashlights. They are chasing *D. ordi* in a single file, their lights flashing wildly over the dunes, and the file makes quick right-angle turns. The van driver muses, wonders why the chasers stay in a single file. There is a sudden convergence and four or five insect nets slap rapidly at the grass. A dusty face appears at the van window, asking for a cage and relaying the tally. It is one of the strangest scenes I have ever witnessed, and I wonder what the guys over their cards in the Colorado rec vehicle would think about it. In the cage *D. ordi* claws the a side and looks out, sometimes jumps only to slam its head against the top. The chase ends at one or two in the morning, and by this time the moon has started to creep above the sandhills to the east.

It is light now, the light of day, and the little girl has put up the flag. Midges have filtered in through the screens and have been ground into wet mush by the early workers in the lab. The cages are in large cardboard cartons, appropriately labeled "very wild animals—do not open." The very wild animals are scratching inside the cages inside the cartons. It is a nocturnal sound, foreign in the morning, out of place with coffee in a plastic cup.

The rats will be combed for chiggers and lice later that day, and the individual parasites will be counted. It is a night thing brought into the day's activities.

The talk at breakfast is of the chase, the stumbling and rolling in the prickly pear, the ones that got away, how Crazy Diane sat in the truck and laughed, how the truck got away and rolled down the hill, all in the dead of night. The cops rub their watery eyes and start down the stairs to work on their rats. The rat is combed and converted into a statistic: number of lice, number of chiggers, and yet months afterward talk will not be of the number of chiggers and lice, but only of the chase. Not even of the catch, only of the chase at night. For a few hours, a few companions, a few insect nets, a few flashlights, and a state van, a few Keystone cops have been allowed to play owl or badger. Playing owl or badger has a way of reminding one of a year past, when Arthur Bay was seined at midnight regularly for several days.

It is a different Arthur Bay at midnight. One comes away from Arthur Bay at midnight with a lingering disturbing feeling that one has not really seen even a fraction of what there is to see in Arthur Bay. More lingering and disturbing is the feeling that maybe one should not learn too much about Arthur Bay, not intrude too much into that time that is Arthur Bay's or even into that time that any person or place should be able to claim for its own, to claim for its own in order to gather strength and thoughts to spend when needed.

It is always a surprise to discover a person or a place is not what one thinks at all, but in reality has a character or several characters apart, withheld from even closest companions, and revealed only when blundered into at times that person or place has rightfully considered its own. Maybe that is why the breakfast talk is only of the chase. The chase is the blunder into Arthur Bay's private time; the data are what was found there when the Keystone cops stumbled into Arthur Bay's private time. The data are very interesting, they tell us much about how *D. ordi* encounters lice and chiggers, but we don't really feel comfortable talking about them to those who have not been along when we stumbled in Arthur Bay's private time.

[193]

As I write this, I wonder if we will go back to Arthur Bay next year. The toads and flies and kangaroo rats and fishermen and rec vehicles will still be there and will always be there even, in a sense, when they're long gone. Maybe it will be Whitetail Creek next year, whose private time we stumble into, whose deer flies remind us so convincingly of our place in the sandhills. It is unscientific for places to become personalities, teachers. I am walking on the highway now, back toward Sportsman's Complex, away from Arthur Bay, and it is a strange, strange feeling. Perhaps it is a time that needs to be considered carefully while one washes test tubes back in the laboratory three hundred miles away. I am not a person and Arthur Bay is not a place; we are similar and Arthur Bay has been my teacher.

It is fall now, the leaves back in the city are turning and the football team is on the front page and it is budget time and senior administrators are resigning and presidential candidates are debating on national television. I never really try to apply those lessons from Arthur Bay. They surface themselves, and come up on the table out front before they can be suppressed. They are often embarrassing. I wonder how a lesson taught so well by nature can be so out of place when applied in human situations. No, we may not go back to Arthur Bay next year or even the year after, but then a fall decision not to return to Arthur Bay in the summer is easily revealed for what it is: a simple case of the Ogallala blues.

13

The Ogallala Blues

S HE CRIED for a month the first time, or maybe it was two months. Openly, that is. She may have cried for almost a year privately. She got up in the late morning, those last summer days, and started her crying. She could give no explanation. She'd started crying out on the highway, out where the dune ridges could be seen beyond the pastures, but then she slept for a while before starting it up again. Sometimes it was sobbing, but more often just a look and silent tears. Even in the city you could not help but think she was seeing those dune ridges out beyond where the cattle were. Her cheeks became raw from the rubbing. Her friends gave her some trouble and she cried, but then she got mad and decided they were wrong and she was right, so there were some angry tears for reasons. Her friends never recovered; she wrote them off, sized them up, and in her mind put them in their place. They could never know how it had been. They were deprived people and perhaps they sensed their deprivation. She had total confidence in her assessment. "Some friends," she'd said. She was in the eighth grade that year, and it was the first documented case of the Ogallala blues.

They see one another on the streets and in the halls. They sometimes smile and they always have a certain look. They share the look with Little Striped Fish. You can spot them a mile away, for they've been to Ogallala and sometimes they're patient with the arrangements of society and sometimes they are not. There are nearly a hundred of them on the streets now. They did things they might not have done elsewhere. The trappings of their lives were seines, the mist net, birds held in the hand, parts of birds and fish seen and realized like the angled rictus of a meadowlark and the dorsal fin of a quillback, midge larvae, a water mite now on a slide in a box on a dark shelf, and the microscopes.

For once the institution had gone first-class and bought quality instruments. For once in its hundred-year-plus-history the institution had provided its students with the ability to truly see, and a termite was no longer a termite but the protozoa in its intestine. Those same microscopes would be used back in the city, of course; and every time, someone who'd been in the field would smile. An old friend in a city lab. What's a nice microscope like you doing in a place like this? It would have to be rigorously demonstrated that one of those microscopes actually refused to focus in the city. Shipping damage, that's the cause of it: an Ogallala microscope would be incapable of purposefully not focusing in the city. So concludes the city professor.

The city professor continues his study of the *Fundulus* chronicles. A trip to the South Platte is made in September, and another in October, late. His doctoral student goes along, and so do a couple of the microscopes. The student has been to Ogallala before, in fact has spent two summers standing in the water of places like the South Platte and Arthur Bay. Eight miles north of Ogallala the microscopes work perfectly. It's taken maybe thirty minutes to seine the fish, drive the eight miles, set up living quarters, build a fire, and begin studying the fish. The microscopes are back on their tables, back home. The student warms his hands in front of the fire, not a sound for miles now that he's back home. They'd stopped in town for a couple of beers in Bill and Rhonda's place, and it had been as if they'd never left, back home. The fish give up the measurements and their gill community unfolds beneath the lenses.

In the city there is a committee meeting. People sit around a table talking about words on a form, and they talk about the words on a form for maybe two hours. The professor is supposed to be there, discussing intelligently. A test is to be given, a test to determine whether the doctoral student can continue his studies, a test to determine whether a defined step in the defined program has been accomplished. The results of the test will go on a form, and the student is supposed to be there, studying. A class is being taught. The subject is one that the teacher has learned from a book that very morning, and it concerns some living things; but no living thing is to be provided the students, although the microscopes are supposed to be there, focusing. Maybe if the work goes quickly, the professor and the student and the microscopes could get back to those responsibilities tomorrow. The work does go quickly, and the student and professor and microscopes do make it back to the city in time, in plenty of time to intelligently discuss words on a form for the next several years, in time to study for a qualifying exam for the next several years, in time to focus for the next several years. Back in the city, they stop in a bar for a couple of beers instead. In the halls the next day a person with a tie will comment on the professor's absence at the committee meeting. The professor will say he's been to Ogallala. In the halls the next week, a person with a tie may comment again on the professor's absence at a committee meeting. Again the professor might just say he's been to Ogallala.

There is work to be done, in this country, before winter. The car must be equipped, the martin house cleaned out and stored, the firewood ordered, the snow tires bought. A person should migrate instead, maybe to Mexico with the cliff swallows. Some time before the firewood was ordered they left. The purple martins left very early, but the barn swallows hung on, at least until night got down into the thirties. They're all gone now. Someplace, maybe around Harlingen, Texas, there are flying insects and swallows this time of year, and those people don't care that they have our swallows and the swallows could not care less that they are in Texas on their way to Mexico and South America. There should be some excuse to migrate. Maybe next year it will be the cliff swallow chronicles, and we'll all fly south in September.

The big question of migration has for centuries been "why," but I find myself asking not "why" but "what do they do down there?" There is no problem reconciling a warbler's or thrush's migration. Their food might change somewhat, but I find it difficult to believe they will be in a microhabitat much different from that lived in in Nebraska. I find it difficult to believe their behavior will change too much; perhaps their song will differ and of course there will be no nesting and courtship. Warblers and thrushes are not colonial, and they are accustomed to hopping around in the branches. No cliff swallow ever hopped around any branch. It is exceedingly difficult to imagine cliff swallows without their nesting colony. Are the birds called a *colony* in South America? After the nesting season they do gather on wires; are there enough wires in South America? One senses *here* that the colony has a territory; will the colony hang together for the migration and establish a territory as a colony in South America, a colony without the nests? One wonders if this time is swallow time, time to gather their strength for the nesting splurge.

In North America we consider migration a burden. Birds prepare metabolically for migration. They carry out physical feats during migration that stretch our imagination. We romanticize migration: in spring there is a welling in the hearts of all these birds that drives them to their nesting grounds to carry out the beautiful business of courtship, species renewal. We further romanticize migration: in fall they assume this metabolic and physical burden, during which all these beautiful birds lose their colors and undertake this trek through untold dangers to a miserable and undeveloped place called South America, so they can escape winter's bitterness in order to breed again. The cliff swallow must be telling us this is all bullshit. Every thing needs its own time. The nesting cliff swallow is locked into the colony's time and tasks. Migration must be a relief that cannot be described. There is nothing noble in staying around to live through the Nebraska winter. There is some merit in taking some time away from the demands of society and calling that time one's own time. Even five minutes does wonders.

"Don't disturb me now."

"You're not doing anything."

"I'm thinking."

"You can think while you —— ."

"No I can't; go away; this is my time to rest my mind and synthesize my approaches."

"You are killing time."

"I am killing time to feed myself the same way we kill cattle to feed ourselves."

Standing in the South Platte beneath the I-80 exit bridge at Ogallala, in late October, one can still gaze up at the cliff swallow nests. Scrambling down the sand bank with seines and buckets brings a flurry of activity from the colony. Waves of birds, even in late October, fall from the nests under the bridge. In October the birds are sparrows. Sparrows are using the cliff swallow colony as a place to roost in the cold. No sparrow ever asked a cliff swallow if the colony could be used. One can sense the cliff swallow's answer: "Sure."

One wonders if the relief from nesting and its concurrent suppression of individuality has returned personality to the individual cliff swallow. Privately *they* must all be wondering at the bludgeoning conservatism that demands a sparrow remain on the Great Plains of North America during January. Right now it's October and the swallows are on their way to a couple of months of their time. In March it will be a different matter. In March the professor will look at a map and discover that the cliff swallows are in Mexico on their way north. It used to be called spring fever. It's now a race to Ogallala, which the cliff swallows will win. In March there is still ice on the windshield, still snow, still driving cold and rain, and "spring break," during which it has never been spring. The race to the spillway is all-consuming. It's another kind of the Ogallala blues, and it blurs the vision and sets the mouth for days and weeks on end. Committee meetings used to be put on the calendar and attended. Now they generally are put on the calendar and forgotten. In March, with the cliff swallows in Mexico headed north, the committee meetings will not even be put on the calendar. There have been hundreds of committee meetings in the past years; hundreds, maybe thousands. There has never been any item of business conducted that was not subsequently changed or overridden. A city professor may not have time for a committee meeting when the cliff swallows are in Mexico headed north.

The trip in October stops temporarily in Gothenburg beneath the microwave tower. An official of the local Public Power and Irrigation District has been contacted, and a side trip will be taken to the North Platte diversion dam. Those prairie rivers come immediately to mind when the diversion dam is first read of. How does one put a diversion dam across the South Platte, or the South Canadian, or the Cimarron? The answer is down the road south of Maxwell and along the irrigation canal: one pours some concrete and makes some metal gates, strings it all out across the sand for a quarter of a mile, then sends the water that backs up down an irrigation canal.

The irrigation canal twists through the sandhills, and some banks are lined with car bodies. The car bodies are not thrown helter-skelter; they are lined up door-to-door neatly along the bank, front ends toward the water. They are all slightly smashed, and slightly smashed to about the same height. The car bodies were obviously very carefully placed in their positions, and some great amounts of care and skill and energy were used to flatten them all about the same. One senses some joy in one's work on the part of the crane operator who placed them so uniformly. Look around from time to time, there may in all of humanity's activities be some traces of an individual's subtle pleasure in what to us is menial work. The crane operator's pleasure has converted the irrigation canal into a piece of modern monumental sculpture. "Irrigation Canal With Car Body"; the sign should be plastic and the title of the sculpture carved in gothic letters; the sign should also be angled slightly so that it can be read easily by the matrons and patrons strolling along the gravel road. The artist needs to be indicated: "Joseph M——, 1931–." The artist would be forty-seven years old this year.

It is bitter cold out on the diversion dam, and the District official is shivering in his light jacket. He had been adding figures with a pocket calculator, in the building beneath the microwave tower at Gothenberg, when we entered the office. He was going to drive us up to the dam, but we followed him in our own car instead. For some reason, it seemed wrong for a couple of guys in jeans to be able to walk into an official's office and get him into his private car to drive up to a diversion dam and irrigation canal forty miles away. It hit us out on the interstate: add a little excitement to someone's life every day. I doubt if we will soon drive to Gothenberg for the sole reason of getting the District official out of his office for an afternoon. Might do it next spring, however, and might also call him in the morning, just so he'll have to cancel all his appointments and notify all his superiors and subordinates. If he took the entire morning canceling appointments and notifying superiors and subordinates, then we might provide him with an entire day of excitement. You can almost picture the official, weighing the phone call from the city against what he knows will go on in a meeting that afternoon. He has a major responsibility in the committee meeting that afternoon; after all, if the committee fails to accomplish its agenda, then there will be nothing for a subsequent committee to alter. The Irrigation District official's mind has manipulated water and rivers and pictures of water and rivers and figures representing water and rivers and canals for days and weeks and months. He cancels his meeting. At some point one must actually leave the imagination of books and go out and stand on the diversion dam: even a PPID official may need to feel the reality of concrete beneath his feet to establish again in his mind the validity of his life with maps and figures. The scientists from the big city are an easy excuse. Everyone in PPID understands.

The official does not understand yet, however. One day in March, although he doesn't know it yet, he will cancel the committee meeting for no particular reason, get in his car and drive to the diversion dam and stand again on the concrete. He will lean on the railing, lighting a cigarette in the wind that whips his light jacket. The dam is a living thing for him, it functions, responds to the giant cables that open the gates, waters the prairie. The dredge behind the dam is also alive for him, and somehow when the dredge is working he senses himself in the presence of a dinosaur. When the dredge is not working, the dinosaur is asleep. He is subconsciously assured that the lines on his wall map do in fact represent real things. He did not actually say this to himself when he walked out past the secretary, cleaning her upper teeth with her tongue, but he felt something urgent and knew that only his own feet on the concrete of the diversion dam would clear his mind.

Those two guys that came by in the fall were carrying a disease. It was contagious and he became infected. They placed value on him and on his time; they placed a certain value on the living diversion dam; they were interested not in the money or cfs flow or water laws or all the rules and regulations that surrounded irrigation by diversion, they just wanted to see the dam and talk about how it worked. In return, they talked about little striped fishes that lived in this river fifty miles west. They were the only people the official had ever talked to who simply wanted to go out on the dam and talk about how it worked without once mentioning money or laws, and they were the only people the official had ever talked to who in return talked only of a little striped fish of no economic importance. They valued the dam because it was there, and they valued the fish because they were there. The District official now values the dam because it is there. He gazes down at the backed-up water and flips the cigarette butt into the river. He wonders if those little fish are swimming around down below the surface. He doesn't want to go back to the office. He has the Ogallala blues.

The irrigation canal was large, large in cross section and wide and long, bending at engineering angles through the Sandhills at the foot of the dune ridges. The dune ridges were not really dune ridges, they just looked it from a distance. Up close, holding off the irrigation canal, they were outright high hills, sculptured and weathered and sharp in their outlines. They were the kind of stuff cowboy and Indian movies are made of, and as we drove along the irrigation canal they moved and shifted and took their places, changing positions and roles, actors in the theatrics of irrigation. The dune ridges out beyond the cattle, the ridges she had been seeing even as she cried back in the city, turned out to be these actor hills. They were farther from the interstate than one realized, actually many miles, and they only looked like dune ridges from that superhighway. Up close, across the irrigation canal, they were carved high bills with juniper-lined canyons between. There was some reason to see these hills, and the feeling that they called for investigation was, up close, fully justified. She had chosen the right thing to see through her tears; and if anyone had asked, there would have been no question in her mind that she had chosen the right thing. She would have known as surely as if she'd been there that these ridges up close would make their moves behind the irrigation canal footlights and tell stories and reveal sights that one could not see from the interstate.

Even an eighth-grader would have known what to do with those ridges and their juniper-lined canyons: set up a mist net across the canyon. There's malaria in them hills. In the car along the gravel irrigation canal road we were both thinking the same thing at exactly the same time. There had to be malaria in those birds that nest in the junipers, and we knew it as surely as we knew that the other is thinking exactly the same conclusion. Not a bit of research had been done in the dune ridges, however, and not one bird's blood had been taken, and not one bird from the irrigation canal place had been sampled and released; but we knew the malaria was there. The hills had told us. It was their way of saying they'd like to be included, that they were as much a part of this land as those canyons eight miles north of Ogallala. This narrative is again getting out of hand: dune ridges can't have the Ogallala blues.

It was the District official who led us here in the first place. He has not mentioned the hills either in his office or standing out on the diversion dam. I wonder if he ever saw them. I also wonder if they might be his private things to look at, and if he thinks that just because we didn't comment on their beauty we had missed them. I wonder how many people he has led past this place knowing these hills were there to be seen and then smiling to himself when those many people never appeared to have seen his hills. He will know in March that those two guys who had stood in the South Platte with their little fish seine had seen those hills. He will feel in March that he's been put on, that those two guys had seen more in his private hills than he had ever seen and they'd seen it because they'd been to Ogallala and furthermore they'd not bothered, standing out on the concrete in October, to share it with him. That realization will simply add to his case of the Ogallala blues.

The District official had already put his snow tires on. The automobile tire has a way of symbolizing obligations that one thinks might perhaps be siphoning off energies best devoted to snails or termites, and the snow tire on top of the average automobile tire symbolizes a dependence in the winter. One can burn wood, put on clothes, build a shelter in the winter time, and these acts seem natural. One must buy snow tires for winter, and this act seems *un*natural. The tires are exceedingly expensive, and I have difficulty imagining a person in possession of a loved vehicle, a surrogate mistress, who could extend that love to the tires. There are those, however, like natural toad-lovers, who spend their lives in tires. A one- or two-hundred-dollar visit to a local tire dealer opens up a part of society one has trouble believing. A person works day-to-day at his job, tries to do the best job he can, reads the newspapers and the news magazines, watches television, keeps informed, skims *Sports Illustrated*, talks politics with friends, and forms an impression of the world and the way humanity lives in it. This impression is shattered by a one- or two-hundred-dollar visit to the local tire dealer.

There are people there who feed their family only by taking orders for new tires. At the tire dealer's garage they seem all absorbed in their work, interested, excited, authoritative. You call about snow tires, and you call a large local tire dealer early in the season. He gives you a price: snow radials for your family station wagon will cost $106 each this year. A person could buy a tape player for his surrogate mistress with the price of only one of those tires. A person could buy not only the tape player but also a couple of Jensen speakers, all for the price of a single snow radial. The tire dealer on the phone assures you he has plenty in stock. It still has not snowed, or become very icy, so the real stimulus to plunk down $212 for two tires has not been felt. For the measly price of two snow radials a guy could buy not only the tape player and a couple of Jensen speakers but also fifteen or sixteen cases of beer. You know what is happening down at the tire dealers. The guy is taking orders, the guys in the shop are mounting and balancing, the family cars are sitting out on the drive with very black new tires with blue-white sidewalls, the daddies are inside using their charge cards. The volume of this activity staggers the mind. You call three weeks later, and they are temporarily out of snow tires. How can a big tire dealer in a *capital* city be out of snow tires? You sense there is something going on that you are not quite a part of. It's almost the same feeling you had when you first picked up a snail.

The student is also buying snow tires this year, but instead of radials his are studded recaps. He is totally organized and has checked the ads and the prices and knows exactly what he wants. He leaves his wheels at his chosen dealer with the instructions to mount and balance two studded recaps of a certain size and call him when the tires are ready. Businesslike businessman doing business with the automobile service business. The days pass, and he asks more frequently now whether the dealer has called. The dealer is a very large company, but has not called. The student's time is very valuable; he is taking the qualifying exam he had no time for out on the South Platte River; he needs his time to study; he has obligations to his animals that he grows weekly in screw-cap test tubes; he has obligations to provide society with information about those animals, even though society doesn't think it needs or wants that information today. He is asking every day now whether the dealer has called. The dealer has not called at the times I have been in the room with our community telephone. He calls the dealer. The dealer has no record of his order. His order had been placed right before a shipment of studded recaps was due in.

"You have no record of my order?"

"Can't find a ticket."

"Well, go ahead and make out another ticket and mount and balance a pair on those wheels I left."

"Sorry, we're out of that size right now."

[205]

"You said you had a big shipment coming in."

"Shipment came in and is all gone now, we're expecting some more very soon. Want us to look for your ticket?"

"You are telling me that you lost my ticket, got in your shipment of studded recaps, and sold the entire shipment of that size, and didn't even call me?"

"You bet! Want us to look for your ticket?"

"If you don't mind. When you find it tear it up." The student has a feeling there is something going on that he is not quite a part of. It's almost the same feeling he had when he first seined up a *Fundulus kansae*. He will spend some time finding another dealer.

In the meantime, a tape player has been installed in my surrogate mistress. The mistress herself has 82,000 miles on her and has been wrecked and gives me static over my tape player. There are no snow tires on either the family station wagon or the student's car with 108,000 miles on it. It is getting close to the dangerous edge of winter now, and it becomes a game to see how close one can live to that edge without paying the snow radial price for not having migrated with the swallows. The question of whether one can afford both a tape player and snow radials comes up. One senses that the question is in the mind of every American who ever found himself or herself in debt, really in debt. Out on the exit ramp bridge over the South Platte River beside the Ogallala Ramada Inn is the answer to that question. Sometime next summer I will pass over that ramp, and my mind will be beaten into submission by the *Best of the Doors* blaring full volume over rear-deck Jensen speakers. The snow radials will have been taken off the family station wagon. By this time the family station wagon may also have a tape player. There will no longer be a question about whether one can afford both snow tires and tape players. Financially one may not be able to afford both. Intellectually one cannot afford only what one needs; intellectually one must also afford what one wants.

Intellectually, if a human affords only what a human needs, then the human puts off forever what the human wants. A human whose wants are always subordinate to his needs may not be a human. Humanity may need wants in order to survive in human form into the next century. There is no need to send a man to the moon or to send an unmanned exploring spaceship to Mars; there is only the want. There is no need to find out why an animal chooses to live on the gills of a killifish. There are really only wants at the edge of our senses. Humanity has a way of assaulting the wanters, of beating them back into their places of proper needing. Humanity does this by converting the wanters' discoveries into the needers' requirements. This circular race almost gives one a rollicking case of the Ogallala blues. The blues are easily subdued; the remedy is on the *Best of the Doors* tape, and it's called "People Are Strange." It's dark as I drive to work now, but in the dark I have located the rewind button and "People Are Strange" pulsates over and over again. There is a magnificent guitar break. It should sound great underneath the South Platte bridge.

"People Are Strange" of course is an easy explanation for some of the things we have experienced in Keith County, but in reality, even though we participate in some circular races, especially where science is concerned, the termite country of Keith County has taught us that people are not so different from the other forms of life on this spaceship. It is hard not to feel an integral part of this planet after discovering that snails pioneer into the wilderness, wrens volunteer for onerous tasks, swallows know of the benefits of cities, and even spring-fed creeks can assume qualities we thought only humans possessed. At philosophical times, reflecting on lessons, a person very often wonders about the forces that brought about this earth and its inhabitants. Since Keith County has taught us we are not so very different after all from snails and swallows, we must wonder whether the plan that brought us here together might not be more general, more all-encompassing, than we realized. Those kinds of thoughts lead to irreverent places, iconoclastic places, nontraditional places, all-encompassing places of the mind. Suddenly my student and I look different to one another, different than we maybe should look if we were to see with the traditions of humanity that have been so established through the recent few years. Thinking philosophically about our close relationships with fish and snails, but standing in the South Platte River staring at one another, we can only question whether these two guys are in fact made in the image of the force that built the planet, whether we are really so unique and different from our animals and our earth.

Humans talk and write things to other humans, but humans don't really communicate well with other animals. But now that we have seen how similar we are to other animals we must wonder whether communications will soon be established. No human has ever analyzed, or really ever *known*, the thoughts of a cliff swallow, so there is no way of knowing whether the swallow might also feel that it was made in the image of the creative force. But a future communication might well reveal that the cliff swallow feels it is unique, blessed with the image of the creator, and so must take care of the human. Thoughts such as these lead to the conclusion that a creator of Keith County probably created relationships and processes, but retained the option of form. That is a scientist's conclusion.

I suppose we should have considered it our good fortune to have thought the scientist's thoughts about potential creative forces sometime before the Creator reappeared. But maybe I imagined it. We *were* somewhat startled, initially, at the form, but we were not at all startled at the values expressed. It *was* rather late in the day, late in a day of hard physical work, so it didn't occur to us immediately that Ogallala would not be an appropriate place for the Lord to reappear on earth. And given our biased idea that the South Platte River was an appropriate place for anything or anybody, and might even be the best place for somebody, the Lord's appearance on the South Platte didn't seem too strange at the time. I even suspect that if we had gone right into town and told some guys in a bar that the Lord had just appeared out by the Texaco station, those guys would have said, "Sure, this is God's country, ain't it?," without even knowing it was also termite country. Yes, I suppose in retrospect it was fortunate that we were standing in the South Platte beneath the exit ramp bridge, the day the Lord reappeared on earth. The Lord's values were pretty obvious from knee-deep in *Fundulus* water, the form we handled with comparative ease, but Her message took some of the wind out of our sails.

She was a very attractive woman in her mid-thirties, with shoulder-length red hair, and She was driving a baby blue 1973 Mustang. When She spoke, the earth trembled and all activity as far as one could see ceased; the wind stopped, the dried grass settled, the sparrows returned to the cliff swallow nests and simply waited rather than scattering like wind-blown leaves. She had pulled off the interstate onto the sandy parking place beside the bridge and from Her car one could hear *The Eagles Their greatest hits*, full volume from the rear-deck speakers.

"Take only what you need," She said, pointing to the seine, "and try a little harder to get all those other minnows back into the water before they die."

We nodded and were suddenly aware of no traffic noise on the interstate or from town.

"You may use those nests for now," She said to the sparrows, "but in the spring when I send the swallows back you will have to find other places."

I could swear the sparrows nodded.

Out on the river bed a lesser yellowlegs lingered and She called to it, "It's late; get going." The yellowlegs skittered across the sand before climbing in full-strength flight toward the south.

"They are beautiful," She said, Her eyes following the bird. "You may seine this river but you will never really understand My world until you follow the yellowlegs."

We were unable to speak in Her presence, but She knew our thoughts and smiled at them. "So you've been to Ogallala and stepped one foot into the world that I made." She leaned against the Mustang and pulled a cigarette from Her shirt pocket, "and you've got the Ogallala blues."

She surveyed the river, the bridge, the giant plastic football player in the distance, the hills beyond the town, but Her eyes were still on the place the yellowlegs had disappeared into the sky. "Does this road go south?"

"The road to Grant? It goes south for at least a ways."

"I was going to Colorado," She said, "but I think I'll follow the yellowlegs this year."

"You're following the yellowlegs on the Grant highway?"

"Yes."

"We could take you to Lewellen instead."

She fished in her jeans pocket for car keys. "I'm going with the yellowlegs this year," She repeated.

"Through Grant?"

"Why not?" The Mustang roared and gravel splattered over the side of the bridge and down the bank as She followed the yellowlegs toward Grant. We'd picked brown snails from the cattails, netted swallows from the spillway, seined fish from the South Platte, walked the Ackley Valley Ranch, and drunk beer in the Sip 'n Sizzle, only to come down with a malignant case of the Ogallala blues, but we had evidently not seen so much of the world of life as we had led ourselves to believe. We stood in the knee-deep water and watched the blue Mustang disappear down the highway just as the yellowlegs had disappeared into the Grant sky.

"I have this feeling there is something going on that we're still not even yet quite a part of," I said.

The student nodded.

OTHER TITLES BY JOHN JANOVY JR.
AVAILABLE IN BISON BOOKS EDITIONS

Back in Keith County
On Becoming a Biologist